www.EffortlessMath.com

... So Much More Online!

✓ FREE Math lessons

✓ More Math learning books!

✓ Mathematics Worksheets

✓ Online Math Tutors

Need a PDF version of this book?

Send email to: Info@EffortlessMath.com

Ace the ISEE Upper Level Math in 30 Days

The Ultimate Crash Course to Beat the ISEE Upper Level Math Test

By

Reza Nazari

& Ava Ross

All inquiries should be addressed to:

info@effortlessMath.com

www.EffortlessMath.com

ISBN–13: 978-1-970036-83-1

ISBN–10: 1-970036-83-4

Published by: Effortless Math Education

www.EffortlessMath.com

Description

The goal of this book is simple. It will help you incorporate the most effective method and the right strategies to prepare for the ISEE Upper Level Math test quickly and effectively.

Ace the ISEE Upper Level Math in 30 Days, which reflects the 2019 test guidelines and topics, is designed to help you hone your math skills, overcome your exam anxiety, and boost your confidence -- and do your best to defeat ISEE Upper Level Math Test. This ISEE Upper Level Math new edition has been updated to replicate questions appearing on the most recent ISEE Upper Level Math tests. This is a precious learning tool for ISEE Upper Level Math test takers who need extra practice in math to improve their ISEE Upper Level Math score. After reviewing this book, you will have solid foundation and adequate practice that is necessary to ace the ISEE Upper Level Math test. **This book is your ticket to ace the ISEE Upper Level Math!**

Ace the ISEE Upper Level Math in 30 Days provides students with the confidence and math skills they need to succeed on the ISEE Upper Level Math, providing a solid foundation of basic Math topics with abundant exercises for each topic. It is designed to address the needs of ISEE Upper Level test takers who must have a working knowledge of basic Math.

Inside the pages of this comprehensive book, students can learn math topics in a structured manner with a complete study program to help them understand essential math skills. It also has many exciting features, including:

- Content 100% aligned with the 2019 ISEE Upper Level test
- Written by ISEE Upper Level Math tutors and test experts
- Complete coverage of all ISEE Upper Level Math concepts and topics which you will be tested
- Step-by-step guide for all ISEE Upper Level Math topics
- Dynamic design and easy-to-follow activities
- Over 2,500 additional ISEE Upper Level math practice questions in both multiple-choice and grid-in formats with answers grouped by topic, so you can focus on your weak areas
- Abundant Math skill building exercises to help test-takers approach different question types that might be unfamiliar to them
- Exercises on different ISEE Upper Level Math topics such as integers, percent, equations, polynomials, exponents and radicals
- 2 full-length practice tests (featuring new question types) with detailed answers

Effortlessly and confidently follow the step-by-step instructions in this book to ace the ISEE Upper Level Math in a short period of time.

ISEE Upper Level Math in 30 Days **is the only book you'll ever need to master Basic Math topics!** It can be used as a self-study course - you do not need to work with a Math tutor. (It can also be used with a Math tutor).

You'll be surprised how fast you master the Math topics covering on ISEE Upper Level Math Test.

Ideal for self-study as well as for classroom usage.

About the Author

Reza Nazari is the author of more than 100 Math learning books including:
– **Math and Critical Thinking Challenges:** For the Middle and High School Student
– **GRE Math in 30 Days**
– **ASVAB Math Workbook 2018 - 2019**
– **Effortless Math Education Workbooks**
– **and many more Mathematics books …**

Reza is also an experienced Math instructor and a test–prep expert who has been tutoring students since 2008. Reza is the founder of Effortless Math Education, a tutoring company that has helped many students raise their standardized test scores—and attend the colleges of their dreams. Reza provides an individualized custom learning plan and the personalized attention that makes a difference in how students view math.

You can contact Reza via email at:
reza@EffortlessMath.com

Find Reza's professional profile at:
goo.gl/zoC9rJ

Contents

Day 1:
Whole Numbers

Math Topics that you'll learn today:

- ✓ Rounding
- ✓ Whole Number Addition and Subtraction
- ✓ Whole Number Multiplication and Division
- ✓ Rounding and Estimates

"If people do not believe that mathematics is simple, it is only because they do not realize how complicated life is."
— John von Neumann

Rounding

Step-by-step guide:

Rounding is putting a number up or down to the nearest whole number or the nearest hundred, etc.

- ✓ *First, find the place value you'll round to.*
- ✓ *Find the digit to the right of the place value you're rounding to. If it is 5 or bigger, add 1 to the place value you're rounding to and put zero for all digits on its right side. If the digit to the right of the place value is less than 5, keep the place value and put zero for all digits to the right.*

Examples:

1) Round 64 to the nearest ten.

The place value of ten is 6. The digit on the right side is 4 (which is less than 5). Keep 6 and put zero for the digit on the right side. The answer is 60. 64 rounded to the nearest ten is 60, because 64 is closer to 60 than to 70.

2) Round 694 to the nearest hundred.

694 rounded to the nearest hundred is 700, because the digit on the right side of hundred place is 9. Add 1 to 6 and put zeros for other digits. The answer is 700.

✍ *Round each number to the nearest ten.*

1) 84 = ____   4) 63 = ____
2) 70 = ____   5) 79 = ____
3) 47 = ____   6) 55 = ____

✍ *Round each number to the nearest hundred.*

7) 185 = ____   10) 109 = ____
8) 254 = ____   11) 222 = ____
9) 729 = ____   12) 311 = ____

Whole Number Addition and Subtraction

Step-by-step guide:

- ✓ *Line up the numbers.*
- ✓ *Start with the unit place. (ones place)*
- ✓ *Regroup if necessary.*
- ✓ *Add or subtract the tens place.*
- ✓ *Continue with other digits.*

Examples:

1) Find the sum. $285 + 145 = ?$

First line up the numbers: $\begin{array}{r} 285 \\ +\,145 \\ \hline \end{array}$ → Start with the unit place. (ones place) $5 + 5 = 10$,

Write 0 for ones place and keep 1, $\begin{array}{r} {\scriptstyle 1} \\ 285 \\ +\,145 \\ \hline 0 \end{array}$, Add the tens place and the digit 1 we kept:

$1 + 8 + 4 = 13$, Write 3 and keep 1, $\begin{array}{r} {\scriptstyle 1\,1} \\ 285 \\ +\,145 \\ \hline 30 \end{array}$

Continue with other digits → $1 + 2 + 1 = 4$ → $\begin{array}{r} {\scriptstyle 1\,1} \\ 285 \\ +\,145 \\ \hline 430 \end{array}$

2) Find the difference. $976 - 453 = ?$

First line up the numbers: $\begin{array}{r} 976 \\ -\,453 \\ \hline \end{array}$, → Start with the unit place. $6 - 3 = 3$, $\begin{array}{r} 976 \\ -\,453 \\ \hline 3 \end{array}$,

Subtract the tens place. $7 - 5 = 2$, $\begin{array}{r} 976 \\ -\,453 \\ \hline 23 \end{array}$, Continue with other digits → $9 - 4 = 5$, $\begin{array}{r} 976 \\ -\,453 \\ \hline 523 \end{array}$

✎ *Find the sum or difference.*

1) $1,264 + 856 =$

2) $2,689 - 456 =$

3) $1,432 - 556 =$

4) $2,820 + 464 =$

5) $2,170 + 245 =$

6) $3,221 + 2,560 =$

7) $3,788 + 1,892 =$

8) $4,238 + 2,576 =$

Whole Number Multiplication

Step-by-step guide:

- ✓ Learn the times tables first! To solve multiplication problems fast, you need to memorize the times table. For example, 3 times 8 is 24 or 8 times 7 is 56.
- ✓ For multiplication, line up the numbers you are multiplying.
- ✓ Start with the ones place and regroup if necessary.
- ✓ Continue with other digits.

Examples:

1) Solve. $500 \times 30 = ?$

Line up the numbers: $\begin{array}{r} 500 \\ \times\,30 \\ \hline \end{array}$, start with the ones place → $0 \times 500 = 0$, $\begin{array}{r} 500 \\ \times\,30 \\ \hline 0 \end{array}$, Continue with other digit which is 3. → $500 \times 3 = 1,500$, $\begin{array}{r} 500 \\ \times\,30 \\ \hline 1,5000 \end{array}$

2) Solve. $820 \times 25 =?$

Line up the numbers: $\begin{array}{r} 820 \\ \times\,25 \\ \hline \end{array}$, start with the ones place → $5 \times 0 = 0$, $\begin{array}{r} 820 \\ \times\,25 \\ \hline 0 \end{array}$, $5 \times 2 = 10$, write 0 and keep 1. $\begin{array}{r} 820 \\ \times\,25 \\ \hline 00 \end{array}$, → $5 \times 8 = 40$, add 1 to 40, the answer is 41. $\begin{array}{r} 820 \\ \times\,25 \\ \hline 4,100 \end{array}$

Now, write 0 in the next line and multiply 820 by 2, using the same process. (Since 2 is in the tens place, we need to write 0 before doing the operation). The answer is 16,400. Add 4,100 and 16,400. The answer is: $4,100 + 16,400 = 20,500$

✎ *Find the missing number.*

1) $24 \times 5 =$ _____
2) $28 \times 8 =$ _____
3) $360 \times 6 =$ _____
4) $500 \times 12 =$ _____

5) $144 \times 51 =$ _____
6) $238 \times 18 =$ _____
7) $362 \times 14 =$ _____
8) $238 \times 28 =$ _____

Whole Number Division

Step-by-step guide:

Division: A typical division problem: Dividend ÷ Divisor = Quotient

- In division, we want to find how many times a number (divisor) is contained in another number (dividend). The result in a division problem is the quotient.
✓ First, write the problem in division format. (dividend is inside; divisor is outside)

$$\text{Divisor}\,\overline{\big)\,\text{Dividend}}^{\text{Quotient}}$$

✓ Now, find how many times divisor goes into dividend. (if it is a big number, break the dividend into smaller numbers by choosing the appropriate number of digits from left. Start from the first digit on the left side of the divided and see if the divisor will go into it. If not, keep moving over one digit to the right in the dividend until you have a number the divisor will go into.
✓ Find number of times the divisor goes into the part of the dividend you chose.
✓ Write the answer above the digit in the dividend you are using and multiply it by the divisor and Write the product under the part of the dividend you are using, then subtract.
✓ Bring down the next digit in the dividend and repeat this process until you have nothing left to bring down.

Example: Solve. $234 \div 4 = ?$

$$4\,\overline{\big)\,234}$$

✓ First, write the problem in division format.
✓ Start from left digit of the dividend. 4 doesn't go into 2. So, choose another digit of the dividend. It is 3.
✓ Now, find how many times 4 goes into 23. The answer is 5.

$$\overset{5}{4\,\overline{\big)\,234}}$$

✓ Write 5 above the dividend part. 4 times 5 is 20. Write 20 below 23 and subtract. The answer is 3.
✓ Now bring down the next digit which is 4. How many times 4 goes into 34? The answer is 8. Write 8 above dividend. This is the final step since there is no other digit of the dividend to bring down. The final answer is 58 and the remainder is 2.

$$\begin{array}{r} 58 \\ 4\,\overline{\big)\,234} \\ -20 \\ \hline 34 \\ -32 \\ \hline 2 \end{array}$$

✎ *Solve.*

1) $240 \div 5 =$ _____
2) $280 \div 4 =$ _____
3) $562 \div 8 =$ _____
4) $522 \div 9 =$ _____

5) $834 \div 8 =$ _____
6) $922 \div 12 =$ _____
7) $880 \div 15 =$ _____
8) $721 \div 22 =$ _____

Rounding and Estimates

Step-by-step guide:

Rounding and estimating are math strategies used for approximating a number. To estimate means to make a rough guess or calculation. To round means to simplify a known number by scaling it slightly up or down.

✓ To estimate a math operation, round the numbers.
✓ For 2-digit numbers, your usually can round to the nearest tens, for 3-digit numbers, round to nearest hundreds, etc.
✓ Find the answer.

Examples:

1) Estimate the sum by rounding each number to the nearest hundred. $153 + 426 =$?
   153 rounded to the nearest hundred is 200. 426 rounded to the nearest hundred is 400.
   Then: $200 + 400 = 600$

2) Estimate the result by rounding each number to the nearest ten. $38 - 26 = $?
   38 rounded to the nearest ten is 40. 25 rounded to the nearest ten is 30.
   Then: $40 - 30 = 10$

✍ *Estimate the sum by rounding each number to the nearest ten.*

1) $14 + 68 = $ _____

2) $82 + 12 = $ _____

3) $43 + 66 = $ _____

4) $47 + 69 = $ _____

5) $553 + 231 = $ _____

6) $418 + 849 = $ _____

✍ *Estimate the product by rounding each number to the nearest ten.*

7) $55 \times 62 = $ _____

8) $14 \times 27 = $ _____

9) $34 \times 66 = $ _____

10) $18 \times 12 = $ _____

11) $62 \times 53 = $ _____

12) $41 \times 26 = $ _____

Answers – Day 1

Rounding

1) 80
2) 70
3) 50
4) 60

5) 80
6) 60
7) 200
8) 300

9) 700
10) 100
11) 200
12) 300

Whole Number Addition and Subtraction

1) 2,120
2) 2,233
3) 876

4) 3,284
5) 2,415
6) 5,781

7) 5,680
8) 6,814

Whole Number Multiplication

1) 120
2) 224
3) 2,160

4) 6,000
5) 7,344
6) 4,284

7) 5,068
8) 6,664

Whole Number Division

1) 48
2) 70
3) $45, r2$

4) 58
5) $104, 2$
6) $153, r4$

7) $58, r10$
8) $32, 17$

Rounding and Estimates

1) 80
2) 90
3) 110
4) 120

5) 780
6) 1,270
7) 3,600
8) 300

9) 2,100
10) 200
11) 3,000
12) 1,200

Day 2:
Fractions

Math Topics that you'll learn today:

- ✓ Comparing Numbers

- ✓ Simplifying Fractions

- ✓ Adding and Subtracting Fractions

- ✓ Multiplying and Dividing Fractions

"A Man is like a fraction whose numerator is what he is and whose denominator is what he thinks of himself. The larger the denominator, the smaller the fraction." ~Tolstoy

Simplifying Fractions

Step-by-step guide:

✓ Evenly divide both the top and bottom of the fraction by $2, 3, 5, 7, \ldots$ etc.

✓ Continue until you can't go any further.

Examples:

1) Simplify $\frac{12}{20}$.

To simplify $\frac{12}{20}$, find a number that both 12 and 20 are divisible by. Both are divisible by 4.

Then: $\frac{12}{20} = \frac{12 \div 4}{20 \div 4} = \frac{3}{5}$

2) Simplify $\frac{64}{80}$.

To simplify $\frac{64}{80}$, find a number that both 64 and 80 are divisible by. Both are divisible by 8 and 16. Then: $\frac{64}{80} = \frac{64 \div 8}{80 \div 8} = \frac{8}{10}$, 8 and 10 are divisible by 2, then: $\frac{8}{10} = \frac{4}{5}$

or $\frac{64}{80} = \frac{64 \div 16}{80 \div 16} = \frac{4}{5}$

✍ *Simplify each fraction.*

1) $\frac{9}{18} =$

2) $\frac{8}{10} =$

3) $\frac{6}{8} =$

4) $\frac{5}{20} =$

5) $\frac{18}{24} =$

6) $\frac{6}{9} =$

7) $\frac{12}{15} =$

8) $\frac{4}{16} =$

9) $\frac{18}{36} =$

10) $\frac{6}{42} =$

11) $\frac{13}{39} =$

12) $\frac{21}{28} =$

Adding and Subtracting Fractions

Step-by-step guide:

- ✓ For "like" fractions (fractions with the same denominator), add or subtract the numerators and write the answer over the common denominator.
- ✓ Find equivalent fractions with the same denominator before you can add or subtract fractions with different denominators.
- ✓ Adding and Subtracting with the same denominator:

$$\frac{a}{b} + \frac{c}{b} = \frac{a+c}{b} \quad , \quad \frac{a}{b} - \frac{c}{b} = \frac{a-c}{b}$$

- ✓ Adding and Subtracting fractions with different denominators:

$$\frac{a}{b} + \frac{c}{d} = \frac{ad+c}{bd} \quad , \quad \frac{a}{b} - \frac{c}{d} = \frac{ad-cb}{bd}$$

Examples:

1) Subtract fractions. $\frac{4}{5} - \frac{3}{5} =$

For "like" fractions, subtract the numerators and write the answer over the common denominator. then: $\frac{4}{5} - \frac{3}{5} = \frac{1}{5}$

2) Subtract fractions. $\frac{2}{3} - \frac{1}{2} =$

For "unlike" fractions, find equivalent fractions with the same denominator before you can add or subtract fractions with different denominators. Use this formula: $\frac{a}{b} - \frac{c}{d} = \frac{ad-cb}{bd}$

$\frac{2}{3} - \frac{1}{2} = \frac{(2)(2)-(1)(3)}{3 \times 2} = \frac{4-3}{6} = \frac{1}{6}$

✎ *Find the sum or difference.*

1) $\frac{1}{3} + \frac{2}{3} =$

4) $\frac{3}{7} + \frac{2}{3} =$

7) $\frac{2}{3} - \frac{1}{6} =$

2) $\frac{1}{2} + \frac{1}{3} =$

5) $\frac{1}{2} - \frac{1}{3} =$

8) $\frac{3}{5} - \frac{1}{2} =$

3) $\frac{2}{5} + \frac{1}{2} =$

6) $\frac{4}{5} - \frac{2}{3} =$

9) $\frac{8}{9} - \frac{2}{5} =$

Multiplying and Dividing Fractions

Step-by-step guide:

- ✓ Multiplying fractions: multiply the top numbers and multiply the bottom numbers.
- ✓ Dividing fractions: Keep, Change, Flip
- ✓ Keep first fraction, change division sign to multiplication, and flip the numerator and denominator of the second fraction. Then, solve!

Examples:

1) Multiplying fractions. $\frac{5}{6} \times \frac{3}{4} =$

Multiply the top numbers and multiply the bottom numbers.

$\frac{5}{6} \times \frac{3}{4} = \frac{5 \times 3}{6 \times 4} = \frac{15}{24}$, simplify: $\frac{15}{24} = \frac{15 \div 3}{24 \div 3} = \frac{5}{8}$

2) Dividing fractions. $\frac{1}{4} \div \frac{2}{3} =$

Keep first fraction, change division sign to multiplication, and flip the numerator and denominator of the second fraction. Then: $\frac{1}{4} \times \frac{3}{2} = \frac{1 \times 3}{4 \times 2} = \frac{3}{8}$

✍ *Find the answers.*

1) $\frac{1}{2} \times \frac{3}{4} =$

2) $\frac{3}{5} \times \frac{2}{3} =$

3) $\frac{1}{4} \times \frac{2}{5} =$

4) $\frac{1}{6} \times \frac{4}{5} =$

5) $\frac{1}{5} \times \frac{1}{4} =$

6) $\frac{2}{5} \times \frac{1}{2} =$

7) $\frac{1}{2} \div \frac{1}{4} =$

8) $\frac{1}{3} \div \frac{1}{2} =$

9) $\frac{2}{5} \div \frac{1}{3} =$

10) $\frac{1}{4} \div \frac{2}{3} =$

11) $\frac{1}{5} \div \frac{3}{10} =$

12) $\frac{2}{7} \div \frac{1}{3} =$

Answers – Day 2

Simplifying Fractions

1) $\frac{1}{2}$

2) $\frac{4}{5}$

3) $\frac{3}{4}$

4) $\frac{1}{4}$

5) $\frac{3}{4}$

6) $\frac{2}{3}$

7) $\frac{4}{5}$

8) $\frac{1}{4}$

9) $\frac{1}{2}$

10) $\frac{1}{7}$

11) $\frac{1}{3}$

12) $\frac{3}{4}$

Adding and Subtracting Fractions

1) $\frac{3}{3} = 1$

2) $\frac{5}{6}$

3) $\frac{9}{10}$

4) $\frac{23}{21}$

5) $\frac{1}{6}$

6) $\frac{2}{15}$

7) $\frac{1}{2}$

8) $\frac{1}{10}$

9) $\frac{22}{45}$

Multiplying and Dividing Fractions

1) $\frac{3}{8}$

2) $\frac{2}{5}$

3) $\frac{1}{10}$

4) $\frac{2}{15}$

5) $\frac{1}{20}$

6) $\frac{1}{5}$

7) 2

8) $\frac{2}{3}$

9) $\frac{6}{5}$

10) $\frac{3}{8}$

11) $\frac{2}{3}$

12) $\frac{6}{7}$

Day 3:
Mixed Numbers

Math Topics that you'll learn today:

✓ Adding Mixed Numbers

✓ Subtracting Mixed Numbers

✓ Multiplying Mixed Numbers

✓ Dividing Mixed Numbers

"Wherever there is number, there is beauty." - Proclus

Adding Mixed Numbers

Step-by-step guide:

Use the following steps for both adding and subtracting mixed numbers.

✓ Add whole numbers of the mixed numbers.
✓ Add the fractions of each mixed number.
✓ Find the Least Common Denominator (LCD) if necessary.
✓ Add whole numbers and fractions.
✓ Write your answer in lowest terms.

Examples:

1) Add mixed numbers. $1\frac{3}{4} + 2\frac{3}{8} =$

Rewriting our equation with parts separated, $1 + \frac{3}{4} + 2 + \frac{3}{8}$, Solving the whole number parts $1 + 2 = 3$, Solving the fraction parts $\frac{3}{4} + \frac{3}{8}$, and rewrite to solve with the equivalent fractions.

$\frac{6}{8} + \frac{3}{8} = \frac{9}{8} = 1\frac{1}{8}$, then Combining the whole and fraction parts $3 + 1 + \frac{1}{8} = 4\frac{1}{8}$

2) Add mixed numbers. $1\frac{2}{3} + 4\frac{1}{6} =$

Rewriting our equation with parts separated, $1 + \frac{2}{3} + 4 + \frac{1}{6}$, Solving the whole number parts $1 + 4 = 5$, Solving the fraction parts $\frac{2}{3} + \frac{1}{6}$, and rewrite to solve with the equivalent fractions.

$\frac{2}{3} + \frac{1}{6} = \frac{5}{6}$, then Combining the whole and fraction parts $5 + \frac{5}{6} = 5\frac{5}{6}$

✎ *Find the sum.*

1) $2\frac{1}{2} + 1\frac{1}{3} =$

2) $6\frac{1}{2} + 3\frac{1}{2} =$

3) $2\frac{3}{8} + 3\frac{1}{8} =$

4) $4\frac{1}{2} + 1\frac{1}{4} =$

5) $1\frac{3}{7} + 1\frac{5}{14} =$

6) $6\frac{5}{12} + 3\frac{3}{4} =$

7) $5\frac{1}{2} + 8\frac{3}{4} =$

8) $3\frac{7}{8} + 3\frac{1}{3} =$

9) $3\frac{3}{9} + 7\frac{6}{11} =$

Subtract Mixed Numbers

Step-by-step guide:

Use the following steps for both adding and subtracting mixed numbers.

- ✓ Subtract the whole number of second mixed number from whole number of the first mixed number.
- ✓ Subtract the second fraction from the first one.
- ✓ Find the Least Common Denominator (LCD) if necessary.
- ✓ Add the result of whole numbers and fractions.
- ✓ Write your answer in lowest terms.

Examples:

1) Subtract. $5\frac{2}{3} - 2\frac{1}{4} =$

Rewriting our equation with parts separated, $5 + \frac{2}{3} - 2 - \frac{1}{4}$

Solving the whole number parts $5 - 2 = 3$, Solving the fraction parts, $\frac{2}{3} - \frac{1}{4} = \frac{8-3}{12} = \frac{5}{12}$

Combining the whole and fraction parts, $3 + \frac{5}{12} = 3\frac{5}{12}$

2) Subtract. $3\frac{4}{5} - 1\frac{1}{2} =$

Rewriting our equation with parts separated, $3 + \frac{4}{5} - 1 - \frac{1}{2}$

Solving the whole number parts $3 - 1 = 2$, Solving the fraction parts, $\frac{4}{5} - \frac{1}{2} = \frac{3}{10}$

Combining the whole and fraction parts, $2 + \frac{3}{10} = 2\frac{3}{10}$

✍ *Find the difference.*

1) $3\frac{1}{3} - 1\frac{1}{3} =$

2) $4\frac{1}{2} - 3\frac{1}{2} =$

3) $5\frac{1}{2} - 2\frac{1}{4} =$

4) $6\frac{1}{6} - 5\frac{1}{3} =$

5) $8\frac{1}{2} - 1\frac{1}{10} =$

6) $9\frac{1}{2} - 2\frac{1}{4} =$

7) $9\frac{1}{5} - 5\frac{1}{6} =$

8) $14\frac{3}{10} - 13\frac{1}{3} =$

9) $19\frac{2}{3} - 11\frac{5}{8} =$

Multiplying Mixed Numbers

Step-by-step guide:

✓ Convert the mixed numbers to improper fractions. (improper fraction is a fraction in which the top number is bigger than bottom number)

✓ Multiply fractions and simplify if necessary.

$$a\frac{c}{b} = a + \frac{c}{b} = \frac{ab + c}{b}$$

Examples:

1) Multiply mixed numbers. $3\frac{2}{3} \times 2\frac{1}{2} =$

Converting mixed numbers to fractions, $3\frac{2}{3} = \frac{11}{3}$ and $2\frac{1}{2} = \frac{5}{2}$.

$\frac{11}{3} \times \frac{5}{2}$, Applying the fractions formula for multiplication, $\frac{11\times5}{3\times2} = \frac{55}{6} = 9\frac{1}{6}$

2) Multiply mixed numbers. $4\frac{3}{5} \times 2\frac{1}{3} =$

Converting mixed numbers to fractions, $\frac{23}{5} \times \frac{7}{3}$, Applying the fractions formula for multiplication, $\frac{23\times7}{5\times3} = \frac{161}{15} = 10\frac{11}{15}$

✎ *Find the product.*

1) $4\frac{1}{3} \times 2\frac{1}{5} =$

2) $3\frac{1}{2} \times 3\frac{1}{4} =$

3) $5\frac{2}{5} \times 2\frac{1}{3} =$

4) $2\frac{1}{2} \times 1\frac{2}{9} =$

5) $3\frac{4}{7} \times 2\frac{3}{5} =$

6) $7\frac{2}{3} \times 2\frac{2}{3} =$

7) $9\frac{8}{9} \times 8\frac{3}{4} =$

8) $2\frac{4}{7} \times 5\frac{2}{9} =$

9) $5\frac{2}{5} \times 2\frac{3}{5} =$

10) $3\frac{5}{7} \times 3\frac{5}{6} =$

Dividing Mixed Numbers

Step-by-step guide:

- ✓ Convert the mixed numbers to improper fractions.

- ✓ Divide fractions and simplify if necessary.

$$a\frac{c}{b} = a + \frac{c}{b} = \frac{ab + c}{b}$$

Examples:

1) Find the quotient. $2\frac{1}{2} \div 1\frac{1}{5} =$

Converting mixed numbers to fractions, $\frac{5}{2} \div \frac{6}{5}$, Applying the fractions formula for multiplication, $\frac{5 \times 5}{2 \times 6} = \frac{25}{12} = 2\frac{1}{12}$

2) Find the quotient. $4\frac{3}{4} \div 3\frac{4}{5} =$

Converting mixed numbers to fractions, $\frac{19}{4} \div \frac{19}{5}$, Applying the fractions formula for multiplication, $\frac{19 \times 5}{4 \times 19} = \frac{95}{76} = 1\frac{1}{4}$

✎ *Find the quotient.*

1) $1\frac{2}{3} \div 3\frac{1}{3} =$

2) $2\frac{1}{4} \div 1\frac{1}{2} =$

3) $10\frac{1}{2} \div 1\frac{2}{3} =$

4) $3\frac{1}{6} \div 4\frac{2}{3} =$

5) $4\frac{1}{8} \div 2\frac{1}{2} =$

6) $2\frac{1}{10} \div 2\frac{3}{5} =$

7) $1\frac{4}{11} \div 1\frac{1}{4} =$

8) $9\frac{1}{2} \div 9\frac{2}{3} =$

9) $8\frac{3}{4} \div 2\frac{2}{5} =$

10) $12\frac{1}{2} \div 9\frac{1}{3} =$

Answers – Day 3

Adding Mixed Numbers

1) $3\frac{5}{6}$

2) 10

3) $5\frac{1}{2}$

4) $5\frac{3}{4}$

5) $2\frac{11}{14}$

6) $10\frac{1}{6}$

7) $14\frac{1}{4}$

8) $7\frac{5}{24}$

9) $10\frac{29}{33}$

Subtract Mixed Numbers

1) 2

2) 1

3) $3\frac{1}{4}$

4) $\frac{5}{6}$

5) $7\frac{2}{5}$

6) $7\frac{1}{4}$

7) $4\frac{1}{30}$

8) $\frac{29}{30}$

9) $8\frac{1}{24}$

Multiplying Mixed Numbers

1) $9\frac{8}{15}$

2) $11\frac{3}{8}$

3) $12\frac{3}{5}$

4) $3\frac{1}{18}$

5) $9\frac{2}{7}$

6) $20\frac{4}{9}$

7) $86\frac{19}{36}$

8) $13\frac{3}{7}$

9) $14\frac{1}{25}$

10) $14\frac{5}{21}$

Dividing Mixed Numbers

1) $\frac{1}{2}$

2) $1\frac{1}{2}$

3) $6\frac{3}{10}$

4) $\frac{19}{28}$

5) $1\frac{13}{20}$

6) $\frac{21}{26}$

7) $1\frac{1}{11}$

8) $\frac{57}{58}$

9) $3\frac{31}{48}$

10) $1\frac{19}{56}$

Day 4:
Decimals

Math Topics that you'll learn today:

- ✓ Comparing Decimals

- ✓ Rounding Decimals

- ✓ Adding and Subtracting Decimals

- ✓ Multiplying and Dividing Decimals

"Do not worry about your difficulties in mathematics. I can assure you mine are still greater." ~ Albert Einstein

Comparing Decimals

Step-by-step guide:

Decimals: is a fraction written in a special form. For example, instead of writing $\frac{1}{2}$ you can write **0.5**.

For comparing decimals:

✓ Compare each digit of two decimals in the same place value.
✓ Start from left. Compare hundreds, tens, ones, tenth, hundredth, etc.
✓ To compare numbers, use these symbols:
- Equal to =,   Less than <,   Greater than >
  Greater than or equal ≥,   Less than or equal ≤

Examples:

1) Compare 0.20 and 0.02.

   0.20 *is greater than* 0.02, because the tenth place of 0.20 is 2, but the tenth place of 0.02 is zero. Then: 0.20 > 0.02

2) Compare 0.0210 and 0.110.

   0.0.110 *is greater than* 0.0210, because the tenth place of 0.110 is 1, but the tenth place of 0.0210 is zero. Then: 0.0210 < 0.110

✎ *Write the correct comparison symbol (>, < or =).*

1) 0.50 ☐ 0.050

2) 0.025 ☐ 0.25

3) 2.060 ☐ 2.07

4) 1.75 ☐ 1.07

5) 4.04 ☐ 0.440

6) 3.05 ☐ 3.5

7) 5.05 ☐ 5.050

8) 1.02 ☐ 1.1

9) 2.45 ☐ 2.125

10) 0.932 ☐ 0.0932

11) 3.15 ☐ 3.150

12) 0.718 ☐ 0.89

Rounding Decimals

Step-by-step guide:

- ✓ We can round decimals to a certain accuracy or number of decimal places. This is used to make calculation easier to do and results easier to understand, when exact values are not too important.
- ✓ First, you'll need to remember your place values: For example:

$$12.4567$$

1: tens   2: ones   4: tenths

5: hundredths   6: thousandths   7: tens thousandths

- ✓ To round a decimal, find the place value you'll round to.
- ✓ Find the digit to the right of the place value you're rounding to. If it is 5 or bigger, add 1 to the place value you're rounding to and remove all digits on its right side. If the digit to the right of the place value is less than 5, keep the place value and remove all digits on the right.

Examples:

1) Round 2.**1837** to the thousandth place value.

First look at the next place value to the right, (tens thousandths). It's 7 and it is greater than 5. Thus add 1 to the digit in the thousandth place.

Thousandth place is 3. $\rightarrow 3 + 1 = 4$, then, the answer is 2.184

2) 2.**1837** rounded to the nearest hundredth.

First look at the next place value to the right of thousandths. It's 3 and it is less than 5, thus remove all the digits to the right. Then, the answer is 2.18.

✍ *Round each decimal to the nearest whole number.*

1) 23.18   3) 14.45   5) 3.95
2) 8.6   4) 7.5   6) 56.7

✍ *Round each decimal to the nearest tenth.*

7) 22.652   9) 47.847   11) 16.184
8) 30.342   10) 82.88   12) 71.79

Adding and Subtracting Decimals

Step-by-step guide:

✓ Line up the numbers.

✓ Add zeros to have same number of digits for both numbers if necessary.

✓ Add or subtract using column addition or subtraction.

Examples:

1) Add. $2.5 + 1.24 =$

First line up the numbers: $\begin{array}{r} 2.5 \\ + 1.24 \\ \hline \end{array}$ → Add zeros to have same number of digits for both

numbers. $\begin{array}{r} 2.50 \\ + 1.24 \\ \hline \end{array}$, Start with the hundredths place. $0 + 4 = 4$, $\begin{array}{r} 2.50 \\ + 1.24 \\ \hline 4 \end{array}$, Continue with tenths

place. $5 + 2 = 7$, $\begin{array}{r} 2.50 \\ + 1.24 \\ \hline .74 \end{array}$. Add the ones place. $2 + 1 = 3$, $\begin{array}{r} 2.50 \\ + 1.24 \\ \hline 3.74 \end{array}$

2) Subtract decimals. $4.67 + 2.15 = \begin{array}{r} 4.67 \\ - 2.15 \\ \hline \end{array}$

Start with the hundredths place. $7 - 5 = 2$, $\begin{array}{r} 4.67 \\ - 2.15 \\ \hline 2 \end{array}$, continue with tenths place. $6 - 1 = 5$

$\begin{array}{r} 4.67 \\ - 2.15 \\ \hline .52 \end{array}$, subtract the ones place. $4 - 2 = 2$, $\begin{array}{r} 4.67 \\ - 2.15 \\ \hline 2.52 \end{array}$.

✎ *Find the sum or difference.*

1) $31.13 - 11.45 =$

2) $35.25 + 24.47 =$

3) $73.50 + 22.78 =$

4) $56.67 - 44.39 =$

5) $71.47 + 16.25 =$

6) $68.99 - 53.61 =$

7) $66.24 - 23.11 =$

8) $39.75 + 12.85 =$

Multiplying and Dividing Decimals

Step-by-step guide:

For Multiplication:

✓ Ignore the decimal point and set up and multiply the numbers as you do with whole numbers.
Count the total number of decimal places in both of the factors.
Place the decimal point in the product.
For Division:

✓ If the divisor is not a whole number, move decimal point to right to make it a whole number. Do the same for dividend.
✓ Divide similar to whole numbers.

Examples:

1) Find the product. $0.50 \times 0.20 =$

Set up and multiply the numbers as you do with whole numbers. Line up the numbers: $\begin{smallmatrix} 50 \\ \times\, 20 \end{smallmatrix}$, Start with

the ones place → $50 \times 0 = 0$, $\begin{smallmatrix} 50 \\ \times 20 \\ \hline 0 \end{smallmatrix}$, Continue with other digits → $50 \times 2 = 100$, $\begin{smallmatrix} 50 \\ \times 20 \\ \hline 1,000 \end{smallmatrix}$, Count the

total number of decimal places in both of the factors. (4). Then Place the decimal point in the product.

Then: $\begin{smallmatrix} 0.50 \\ \times\, 0.20 \\ \hline 0.1000 \end{smallmatrix}$ → $0.50 \times 0.20 = 0.1$

2) Find the quotient. $1.20 \div 0.2 =$

The divisor is not a whole number. Multiply it by 10 to get 2. Do the same for the dividend to get 12. Now, divide: $12 \div 2 = 6$. The answer is 6.

✍ *Find the product and quotient.*

1) $0.5 \times 0.4 =$	5) $1.92 \times 0.8 =$	9) $4.2 \div 2 =$
2) $2.5 \times 0.2 =$	6) $0.55 \times 0.4 =$	10) $8.6 \div 0.5 =$
3) $1.25 \times 0.5 =$	7) $1.67 \div 100 =$	11) $12.6 \div 0.2 =$
4) $0.75 \times 0.2 =$	8) $52.2 \div 1,000 =$	12) $16.5 \div 5 =$

Answers – Day 4

Comparing Decimals

1) >
2) <
3) <
4) >
5) >
6) <

7) =
8) <
9) >
10) >
11) =
12) <

Rounding Decimals

1) 23
2) 9
3) 14
4) 8

5) 4
6) 57
7) 22.7
8) 30.3

9) 47.8
10) 82.9
11) 16.2
12) 71.8

Adding and Subtracting Decimals

1) 19.68
2) 59.72
3) 96.28

4) 12.28
5) 87.72
6) 15.38

7) 43.13
8) 52.60

Multiplying and Dividing Decimals

1) 0.20
2) 0.50
3) 0.625
4) 0.15

5) 1.536
6) 0.22
7) 0.0167
8) 0.0522

9) 2.10
10) 17.2
11) 63
12) 3.3

Day 5:
Factoring Numbers

Math Topics that you'll learn today:

- ✓ Factoring Numbers

- ✓ Greatest Common Factor

- ✓ Least Common Multiple

"The study of mathematics, like the Nile, begins in minuteness but ends in magnificence."

~ Charles Caleb Colton

Factoring Numbers

Step-by-step guide:

- ✓ Factoring numbers means to break the numbers into their prime factors.
- ✓ First few prime numbers: $2, 3, 5, 7, 11, 13, 17, 19$

Examples:

1) List all positive factors of 12.

   Write the upside-down division:
   The second column is the answer.
   Then: $12 = 2 \times 2 \times 3$ or $12 = 2^2 \times 3$

12	2
6	2
3	3
1	

2) List all positive factors of 20.

   Write the upside-down division:
   The second column is the answer.
   Then: $20 = 2 \times 2 \times 5$ or $20 = 2^2 \times 5$

20	2
10	2
5	5
1	

✎ *List all positive factors of each number.*

1) 8	5) 25	9) 42
2) 9	6) 28	10) 48
3) 15	7) 26	11) 50
4) 16	8) 35	12) 36

Greatest Common Factor

Step-by-step guide:

- ✓ List the prime factors of each number.
- ✓ Multiply common prime factors.
- ✓ If there are no common prime factors, the GCF is 1.

Examples:

1) Find the GCF for 10 and 15.

   The factors of 10 are: $\{1, 2, 5, 10\}$

   The factors of 15 are: $\{1, 3, 5, 15\}$

   There is 5 in common,

   Then the greatest common factor is: 5.

2) Find the GCF for 8 and 20.

   The factors of 8 are: $\{1, 2, 4, 8\}$

   The factors of 20 are: $\{1, 2, 4, 5, 10, 20\}$

   There is 2 and 4 in common.

   Then the greatest common factor is: $2 \times 4 = 8$.

✍ *Find the GCF for each number pair.*

1) 4, 2	5) 5, 10	9) 5, 12
2) 3, 5	6) 6, 12	10) 4, 14
3) 2, 6	7) 7, 14	11) 15, 18
4) 4, 7	8) 6, 14	12) 12, 20

Least Common Multiple

Step-by-step guide:

- ✓ Least Common Multiple is the smallest multiple that 2 or more numbers have in common.
- ✓ How to find LCM: list out all the multiples of each number and then find the first one they have in common,

Examples:

1) Find the LCM for 3 and 4.

   Multiples of 3: $3, 6, 9, 12, 15, 18, 21, 24$

   Multiples of 4: $4, 8, 12, 16, 20, 24$

   $LCM = 12$

2) Find the LCM for 9 and 12.

   Multiples of 9: $9, 18, 27, 36, 45$

   Multiples of 12: $12, 24, 36, 48$

   $LCM = 36$

✎ *Find the LCM for each number pair.*

1) $3, 6$	5) $6, 18$	9) $4, 18$
2) $5, 10$	6) $10, 12$	10) $9, 12$
3) $6, 14$	7) $4, 12$	11) $12, 16$
4) $8, 9$	8) $5, 15$	12) $15, 18$

Answers – Day 5

Factoring Numbers

1) $1, 2, 4, 8$
2) $1, 3, 9$
3) $1, 3, 5, 15$
4) $1, 2, 4, 8, 16$
5) $1, 5, 25$
6) $1, 2, 4, 7, 14, 28$

7) $1, 2, 13, 26$
8) $1, 5, 7, 35$
9) $1, 2, 3, 6, 7, 14, 21, 42$
10) $1, 2, 3, 4, 6, 8, 12, 16, 24, 48$
11) $1, 2, 5, 10, 25, 50$
12) $1, 2, 3, 4, 6, 9, 12, 18, 36$

Greatest Common Factor

1) 2
2) 1
3) 2
4) 1
5) 5
6) 12

7) 7
8) 2
9) 1
10) 2
11) 3
12) 8

Least Common Multiple

1) 6
2) 10
3) 42
4) 72
5) 18
6) 60

7) 12
8) 15
9) 36
10) 36
11) 48
12) 90

Day 6:

Integers

Math Topics that you'll learn today:

- ✓ Adding and Subtracting Integers

- ✓ Multiplying and Dividing Integers

- ✓ Ordering Integers and Numbers

"Without mathematics, there's nothing you can do. Everything around you is mathematics. Everything around you is numbers." - Shakuntala Devi

Adding and Subtracting Integers

Step-by-step guide:

- ✓ Integers includes: zero, counting numbers, and the negative of the counting numbers. $\{\ldots, -3, -2, -1, 0, 1, 2, 3, \ldots\}$
- ✓ Add a positive integer by moving to the right on the number line.
- ✓ Add a negative integer by moving to the left on the number line.
- ✓ Subtract an integer by adding its opposite.

Examples:

1) Solve. $(-8) - (-5) =$

Keep the first number, and convert the sign of the second number to it's opposite. (change subtraction into addition. Then: $(-8) + 5 = -3$

2) Solve. $10 + (4 - 8) =$

First subtract the numbers in brackets, $4 - 8 = -4$

Then: $10 + (-4) = \;\rightarrow$ change addition into subtraction: $10 - 4 = 6$

✎ *Find each sum or difference.*

1) $12 + (-5) =$

2) $(-14) + (-18) =$

3) $8 + (-28) =$

4) $43 + (-12) =$

5) $(-7) + (-11) + 4 =$

6) $37 + (-16) + 12 =$

7) $(-12) - (-8) =$

8) $15 - (-20) =$

9) $(-11) - 25 =$

10) $30 - (-16) =$

11) $56 - (45 - 23) =$

12) $15 - (-4) - (-34) =$

Multiplying and Dividing Integers

Step-by-step guide:

Use these rules for multiplying and dividing integers:
- ✓ (negative) × (negative) = positive
- ✓ (negative) ÷ (negative) = positive
- ✓ (negative) × (positive) = negative
- ✓ (negative) ÷ (positive) = negative
- ✓ (positive) × (positive) = positive

Examples:

1) Solve. $(2 - 5) \times (3) =$

First subtract the numbers in brackets, $2 - 5 = -3 \rightarrow (-3) \times (3) =$

Now use this formula: (negative) × (positive) = negative
$(-3) \times (3) = -9$

2) Solve. $(-12) + (48 \div 6) =$

First divided 48 by 6 , the numbers in brackets, $48 \div 6 = 8$

$= (-12) + (8) = -12 + 8 = -4$

✎ *Find each product or quotient.*

1) $(-7) \times (-8) =$

2) $(-4) \times 5 =$

3) $5 \times (-11) =$

4) $(-5) \times (-20) =$

5) $-(2) \times (-8) \times 3 =$

6) $(12 - 4) \times (-10) =$

7) $16 \div (-4) =$

8) $(-25) \div (-5) =$

9) $(-40) \div (-8) =$

10) $64 \div (-8) =$

11) $(-49) \div 7 =$

12) $(-112) \div (-4) =$

Ordering Integers and Numbers

Step-by-step guide:

- ✓ When using a number line, numbers increase as you move to the right.
- ✓ When comparing two numbers, think about their position on number line. If one number is on the right side of another number, it is a bigger number. For example, -3 is bigger than -5 because it is on the right side of -5 on number line.

Examples:

1) Order this set of integers from least to greatest. $-2, 1, -5, -1, 2, 4$
   The smallest number is -5 and the largest number is 4.

   Now compare the integers and order them from greatest to least:
   $-5 < -2 < -1 < 1 < 2 < 4$

2) Order each set of integers from greatest to least. $10, -6, -2, 5, -8, 4$
   The largest number is 10 and the smallest number is -8.

   Now compare the integers and order them from least to greatest:
   $10 > 5 > 4 > -2 > -6 > -8$

✎ *Order each set of integers from least to greatest.*

1) $7, -9, -6, -1, 3$ ___, ___, ___, ___, ___, ___
2) $-4, -11, 5, 12, 9$ ___, ___, ___, ___, ___, ___
3) $18, -12, -19, 21, -20$ ___, ___, ___, ___, ___, ___
4) $-15, -25, 18, -7, 32$ ___, ___, ___, ___, ___, ___

✎ *Order each set of integers from greatest to least.*

5) $11, 16, -9, -12, -4$ ___, ___, ___, ___, ___, ___
6) $23, 31, -14, -20, 39$ ___, ___, ___, ___, ___, ___
7) $45, -21, -18, 55, -5$ ___, ___, ___, ___, ___, ___
8) $68, 81, -14, -10, 94$ ___, ___, ___, ___, ___, ___

Answers – Day 6

Adding and Subtracting Integers

1) 7
2) −32
3) −20
4) 31
5) −14
6) 33

7) −4
8) 35
9) −36
10) 46
11) 34
12) 53

Multiplying and Dividing Integers

1) 56
2) −20
3) −55
4) 100
5) 48
6) −80

7) −4
8) 5
9) 5
10) −8
11) −7
12) 28

Ordering Integers and Numbers

1) −9, −6, −1, 3, 7
2) −11, −4, 5, 9, 12
3) −20, −19, −12, 18, 21
4) −25, −15, −7, 18, 32

5) 16, 11, −4, −9, −12
6) 39, 31, 23, −14, −20
7) 55, 45, −5, −18, −21
8) 94, 81, 68, −10, −14

Day 7:
Order of Operations

Math Topics that you'll learn today:

✓ Order of Operations

✓ Integers and Absolute Value

"Sometimes the questions are complicated and the answers are simple." - Dr. Seuss

Order of Operations

Step-by-step guide:

When there is more than one math operation, use PEMDAS:

✓ Parentheses

✓ Exponents

✓ Muliplication and Division (from left to right)

✓ Addition and Subtraction (from left to right)

Examples:

1) Solve. $(5 + 7) \div (3^2 \div 3) =$

First simplify inside parentheses: $(12) \div (9 \div 3) = (12) \div (3) =$
Then: $(12) \div (3) = 4$

2) Solve. $(11 \times 5) - (12 - 7) =$

First simplify inside parentheses: $(11 \times 5) - (12 - 7) = (55) - (5) =$

Then: $(55) - (5) = 50$

✒ *Evaluate each expression.*

1) $5 + (4 \times 2) =$

2) $13 - (2 \times 5) =$

3) $(16 \times 2) + 18 =$

4) $(12 - 5) - (4 \times 3) =$

5) $25 + (14 \div 2) =$

6) $(18 \times 5) \div 5 =$

7) $(48 \div 2) \times (-4) =$

8) $(7 \times 5) + (25 - 12) =$

9) $64 + (3 \times 2) + 8 =$

10) $(20 \times 5) \div (4 + 1) =$

11) $(-9) + (12 \times 6) + 15 =$

12) $(7 \times 8) - (56 \div 4) =$

Integers and Absolute Value

Step-by-step guide:

✓ To find an absolute value of a number, just find its distance from 0 on number line! For example, the distance of 12 and -12 from zero on number line is 12!

Examples:

1) Solve. $\frac{|-18|}{9} \times |5 - 8| =$

First find $|-18|$, →the absolute value of -18 is 18, then: $|-18| = 18$

$\frac{18}{9} \times |5 - 8| =$

Next, solve $|5 - 8|$, → $|5 - 8| = |-3|$, the absolute value of -3 is 3. $|-3| = 3$

Then: $\frac{18}{9} \times 3 = 2 \times 3 = 6$

2) Solve. $|10 - 5| \times \frac{|-2 \times 6|}{3} =$

First solve $|10 - 5|$, → $|10 - 5| = |5|$, the absolute value of 5 is 5, $|5| = 5$

$5 \times \frac{|-2 \times 6|}{3} =$

Now solve $|-2 \times 6|$, → $|-2 \times 6| = |-12|$, the absolute value of -12 is 12, $|-12| = 12$

Then: $5 \times \frac{12}{3} = 5 \times 4 = 20$

✍ *Evaluate the value.*

1) $8 - |2 - 14| - |-2| =$

2) $|-2| - \frac{|-1|}{2} =$

3) $\frac{|-3|}{6} \times |-6| =$

4) $\frac{|5 \times -3|}{5} \times \frac{|-20|}{4} =$

5) $|2 \times -4| + \frac{|-40|}{5} =$

6) $\frac{|-28|}{4} \times \frac{|-55|}{11} =$

7) $|-12 + 4| \times \frac{|-4 \times 5|}{2} =$

8) $\frac{|-10 \times 3|}{2} \times |-12| =$

Answers – Day 7

Order of Operations

1) 13
2) 3
3) 50
4) −5
5) 32
6) 18

7) −96
8) 48
9) 78
10) 20
11) 78
12) 42

Integers and Absolute Value

1) −6
2) −3
3) 36
4) 15

5) 16
6) 35
7) 80
8) 180

Day 8:
Ratios

Math Topics that you'll learn today:

- ✓ Simplifying Ratios

- ✓ Proportional Ratios

Mathematics is the door and key to the sciences. ~ Roger Bacon

Simplifying Ratios

Step-by-step guide:

- ✓ Ratios are used to make comparisons between two numbers.
- ✓ Ratios can be written as a fraction, using the word "to", or with a colon.
- ✓ You can calculate equivalent ratios by multiplying or dividing both sides of the ratio by the same number.

Examples:

1) Simplify. $8 : 4 =$

Both numbers 8 and 4 are divisible by 4 , $\Rightarrow 8 \div 4 = 2, 4 \div 4 = 1,$

Then: $8 : 4 = 2 : 1$

2) Simplify. $\frac{12}{36} =$

Both numbers 12 and 36 are divisible by 12, $\Rightarrow 12 \div 12 = 1, 36 \div 12 = 3,$

Then: $\frac{12}{36} = \frac{1}{3}$

✎ *Reduce each ratio.*

1) $12 : 8 = $ ___ : ___

2) $2 : 20 = $ ___ : ___

3) $3 : 36 = $ ___ : ___

4) $8 : 16 = $ ___ : ___

5) $6 : 100 = $ ___ : ___

6) $10 : 60 = $ ___ : ___

7) $21 : 49 = $ ___ : ___

8) $20 : 40 = $ ___ : ___

9) $10 : 50 = $ ___ : ___

10) $14 : 18 = $ ___ : ___

11) $45 : 27 = $ ___ : ___

12) $49 : 21 = $ ___ : ___

Proportional Ratios

Step-by-step guide:

- ✓ A proportion means that two ratios are equal. It can be written in two ways:
  $\frac{a}{b} = \frac{c}{d}$, $a : b = c : d$
- ✓ The proportion $\frac{a}{b} = \frac{c}{d}$ can be written as: $a \times d = c \times b$

Examples:

1) Solve this proportion for x. $\frac{4}{8} = \frac{5}{x}$

   Use cross multiplication: $\frac{4}{8} = \frac{5}{x} \Rightarrow 4 \times x = 5 \times 8 \Rightarrow 4x = 40$

   Divide to find x: $x = \frac{40}{4} \Rightarrow x = 10$

2) If a box contains red and blue balls in ratio of $2 : 3$ red to blue, how many red balls are there if 90 blue balls are in the box?

   Write a proportion and solve. $\frac{2}{3} = \frac{x}{90}$

   Use cross multiplication: $2 \times 90 = 3 \times x \Rightarrow 180 = 3x$

   Divide to find x: $x = \frac{180}{3} \Rightarrow x = 60$

✍ *Solve each proportion.*

1) $\frac{2}{5} = \frac{14}{x}, x = $ _____

2) $\frac{1}{6} = \frac{7}{x}, x = $ _____

3) $\frac{3}{5} = \frac{27}{x}, x = $ _____

4) $\frac{1}{5} = \frac{x}{80}, x = $ _____

5) $\frac{3}{7} = \frac{x}{63}, x = $ _____

6) $\frac{1}{4} = \frac{13}{x}, x = $ _____

7) $\frac{7}{9} = \frac{56}{x}, x = $ _____

8) $\frac{6}{11} = \frac{42}{x}, x = $ _____

9) $\frac{4}{7} = \frac{x}{77}, x = $ _____

10) $\frac{5}{13} = \frac{x}{143}, x = $ _____

11) $\frac{7}{19} = \frac{x}{209}, x = $ _____

12) $\frac{3}{13} = \frac{x}{195}, x = $ _____

Answers – Day 8

Simplifying Ratios

1) $3:2$
2) $1:10$
3) $1:12$
4) $1:2$
5) $3:50$
6) $1:6$

7) $3:7$
8) $1:2$
9) $1:5$
10) $7:9$
11) $5:3$
12) $7:3$

Proportional Ratios

1) 35
2) 42
3) 45
4) 16
5) 27
6) 52

7) 72
8) 77
9) 44
10) 55
11) 77
12) 45

Day 9:
Similarity and Proportions

Math Topics that you'll learn today:

✓ Create a Proportion

✓ Similarity and Ratios

✓ Simple Interest

Mathematics – the unshaken Foundation of Sciences, and the plentiful Fountain of Advantage to human affairs. –

Isaac Barrow

Create a Proportion

Step-by-step guide:

- ✓ A proportion contains two equal fractions! A proportion simply means that two fractions are equal.
- ✓ To create a proportion, simply find (or create) two equal fractions.

Examples:

1) Express ratios as a Proportion.
120 miles on 4 gallons of gas, how many miles on 1 gallon of gas?

First create a fraction: $\frac{120 \ miles}{4 \ gallons}$, and divide: $120 \div 4 = 30$

Then: 30 miles per gallon

2) State if this pair of ratios form a proportion. $\frac{3}{5} \ and \ \frac{24}{45}$

Use cross multiplication: $\frac{3}{5} = \frac{24}{45} \rightarrow 3 \times 45 = 5 \times 24 \rightarrow 135 = 120$, which is not correct. Therefore, this pair of ratios doesn't form a proportion.

✎ *State if each pair of ratios form a proportion.*

1) $\frac{3}{10} \ and \ \frac{9}{30}$

2) $\frac{1}{2} \ and \ \frac{16}{32}$

3) $\frac{5}{6} \ and \ \frac{35}{42}$

4) $\frac{3}{7} \ and \ \frac{27}{72}$

5) $\frac{2}{5} \ and \ \frac{16}{45}$

6) $\frac{4}{9} \ and \ \frac{40}{81}$

7) $\frac{6}{11} \ and \ \frac{42}{77}$

8) $\frac{1}{6} \ and \ \frac{8}{48}$

9) $\frac{6}{17} \ and \ \frac{36}{85}$

10) $\frac{2}{7} \ and \ \frac{24}{86}$

11) $\frac{12}{19} \ and \ \frac{156}{247}$

12) $\frac{13}{21} \ and \ \frac{182}{294}$

Similarity and Ratios

Step-by-step guide:

✓ Two or more figures are similar if the corresponding angles are equal, and the corresponding sides are in proportion.

Examples:

1) A girl $160\ cm$ tall, stands $360\ cm$ from a lamp post at night. Her shadow from the light is $90\ cm$ long. How high is the lamp post?

Write the proportion and solve for missing side.

$$\frac{\text{Smaller triangle height}}{\text{Smaller triangle base}} = \frac{\text{Bigger triangle height}}{\text{Bigger triangle base}}$$

$$\Rightarrow \frac{90cm}{160cm} = \frac{90+360cm}{x} \Rightarrow 90x = 160 \times 450 \Rightarrow x = 800\ cm$$

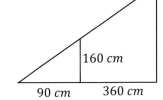

2) A tree $32\ feet$ tall casts a shadow $12\ feet$ long. Jack is $6\ feet$ tall. How long is Jack's shadow?

Write a proportion and solve for the missing number.

$$\frac{32}{12} = \frac{6}{x} \rightarrow 32x = 6 \times 12 = 72$$

$$32x = 72 \rightarrow x = \frac{72}{32} = 2.25$$

✍ *Solve.*

1) Two rectangles are similar. The first is $6\ feet$ wide and $20\ feet$ long. The second is $15\ feet$ wide. What is the length of the second rectangle? _____

2) Two rectangles are similar. One is $2.5\ meters$ by $9\ meters$. The longer side of the second rectangle is $22.5\ meters$. What is the other side of the second rectangle? _____

3) A building casts a shadow $24\ ft$ long. At the same time a girl $5\ ft$ tall casts a shadow $2\ ft$ long. How tall is the building? _____

4) The scale of a map of Texas is $2\ inches: 45\ miles$. If you measure the distance from Dallas to Martin County as $14.4\ inches$, approximately how far is Martin County from Dallas? _____

Simple Interest

Step-by-step guide:

✓ Simple Interest: The charge for borrowing money or the return for lending it. To solve a simple interest problem, use this formula:

Interest = principal × rate × time ⇒ $I = p \times r \times t$

Examples:

1) Find simple interest for $450 investment at 7% for 8 years.

Use Interest formula: $I = prt$

$P = \$450$, r = 7% = $\frac{7}{100}$ = 0.07 and $t = 8$

Then: $I = 450 \times 0.07 \times 8 = \$252$

2) Find simple interest for $5,200 at 4% for 3 years.

Use Interest formula: $I = prt$

$P = \$5,200$, r = 4% = $\frac{4}{100}$ = 0.04 and $t = 3$

Then: $I = 5,200 \times 0.04 \times 3 = \$624$

✎ *Determine the simple interest for these loans.*

1) $1,300 at 5% for 6 years. $ _____

2) $5,400 at 3.5% for 6 months. $ _____

3) $600 at 4% for 9 months. $ _____

4) $24,000 at 5.5% for 5 years. $ _____

5) $15,600 at 3% for 2 years. $ _____

6) $1,200 at 5.5% for 4 years. $ _____

7) $1,600 at 4.5% for 9 months. $ _____

8) $12,000 at 2.2% for 5 years. $ _____

Answers – Day 9

Create a Proportion

1) Yes
2) Yes
3) Yes
4) No
5) No
6) No

7) Yes
8) Yes
9) No
10) No
11) Yes
12) Yes

Similarity and ratios

1) 50 feet
2) 6.25 meters
3) 60 feet
4) 324 miles

Simple Interest

1) $390.00
2) $94.50
3) $18.00
4) $6,600.00

5) $936.00
6) $264.00
7) $54
8) $1,320.00

Day 10:
Percentage

Math Topics that you'll learn today:

✓ Percentage Calculations

✓ Percent Problems

Mathematics is no more computation than typing is literature.

- John Allen Paulos

Percentage Calculations

Step-by-step guide:

- ✓ Percent is a ratio of a number and 100. It always has the same denominator, 100. Percent symbol is %.
- ✓ Percent is another way to write decimals or fractions. For example:

$$40\% = 0.40 = \frac{40}{100} = \frac{2}{5}$$

- ✓ Use the following formula to find part, whole, or percent:

$$\text{part} = \frac{\text{percent}}{100} \times \text{whole}$$

Examples:

1) What is 10% of 45? Use the following formula: $\text{part} = \frac{\text{percent}}{100} \times \text{whole}$

$$\text{part} = \frac{10}{100} \times 45 \;\rightarrow\; \text{part} = \frac{1}{10} \times 45 \rightarrow \text{part} = \frac{45}{10} \rightarrow \text{part} = 4.5$$

2) What is 15% of 24? Use the percent formula: $\text{part} = \frac{\text{percent}}{100} \times \text{whole}$

$$\text{part} = \frac{15}{100} \times 24 \;\rightarrow\; \text{part} = \frac{360}{100} \;\rightarrow\; \text{part} = 3.6$$

✎ *Calculate the given percent of each value.*

1) 2% of 50 = _____

2) 10% of 30 = _____

3) 20% of 25 = _____

4) 50% of 80 = _____

5) 40% of 200 = _____

6) 20% of 45 = _____

7) 35% of 20 = _____

8) 12% of 400 = _____

9) 40% of 90 = _____

10) 25% of 812 = _____

11) 32% of 600 = _____

12) 87% of 500 = _____

Percent Problems

Step-by-step guide:

✓ In each percent problem, we are looking for the base, or part or the percent.
✓ Use the following equations to find each missing section.
  ○ Base = Part ÷ Percent
  ○ Part = Percent × Base
  ○ Percent = Part ÷ Base

Examples:

1) 1.2 is what percent of 24?

In this problem, we are looking for the percent. Use the following equation:
$$Percent = Part \div Base \rightarrow Percent = 1.2 \div 24 = 0.05 = 5\%$$

2) 20 is 5% of what number?

Use the following formula: $Base = Part \div Percent \rightarrow Base = 20 \div 0.05 = 400$
20 is 5% of 400.

✎ *Solve each problem.*

1) 20 is what percent of 50? ____%

2) 18 is what percent of 90? ____%

3) 12 is what percent of 15? ____%

4) 16 is what percent of 200? ____%

5) 24 is what percent of 800? ____%

6) 48 is what percent of 4,00? ____%

7) 90 is 12 percent of what number? ____

8) 24 is 8 percent of what? ____

9) 60 is 15 percent of what number? ____

10) 42 *is 12 percent of what?* ___

11) 11 *is 25 percent of what?* ___

12) 8 *is 12.5 percent of what?* ___

Answers – Day 10

Percentage Calculations

1) 1
2) 3
3) 5
4) 40
5) 80
6) 9

7) 7
8) 48
9) 36
10) 203
11) 192
12) 435

Percent Problems

1) 40%
2) 20%
3) 80%
4) 8%
5) 3%
6) 12%

7) 750
8) 300
9) 400
10) 350
11) 44
12) 64

Day 11:
Percent of Change

Math Topics that you'll learn today:

✓ Percent of Increase and Decrease

✓ Discount, Tax and Tip

Mathematics is a great motivator for all humans. Because its career starts with zero and it never end (infinity).

Percent of Increase and Decrease

Step-by-step guide:

To find the percentage of increase or decrease:
- ✓ New Number – Original Number
- ✓ The result ÷ Original Number × 100
- ✓ If your answer is a negative number, then this is a percentage decrease. If it is positive, then this is a percent of increase.

Examples:

1) Increased by 50%, the numbers 84 becomes:

   First find 50% of 84 → $\frac{50}{100} \times 84 = \frac{50 \times 84}{100} = 42$

   Then: $84 + 42 = 126$

2) The price of a shirt increases from $10 to $14. What is the percent increase?
   First: $14 - 10 = 4$
   4 is the result. Then: $4 \div 10 = \frac{4}{10} = 0.4 = 40\%$

✎ *Solve each percent of change word problem.*

1) Bob got a raise, and his hourly wage increased from $12 to $15. What is the percent increase? _____ %

2) The price of a pair of shoes increases from $20 to $32. What is the percent increase? ____ %

3) At a coffeeshop, the price of a cup of coffee increased from $1.20 to $1.44. What is the percent increase in the cost of the coffee? _____ %

4) 6 *cm* are cut from a 24 *cm* board. What is the percent decrease in length? _____ %

5) In a class, the number of students has been increased from 18 to 27. What is the percent increase? _____ %

6) The price of gasoline rose from $2.40 to $2.76 in one month. By what percent did the gas price rise? _____ %

7) A shirt was originally priced at $48. It went on sale for $38.40. What was the percent that the shirt was discounted? _____ %

Discount, Tax and Tip

Step-by-step guide:

- ✓ Discount = Multiply the regular price by the rate of discount
- ✓ Selling price = original price – discount
- ✓ Tax: To find tax, multiply the tax rate to the taxable amount (income, property value, etc.)
- ✓ To find tip, multiply the rate to the selling price.

Examples:

1) With an 10% discount, Ella was able to save $20 on a dress. What was the original price of the dress?

$10\% \ of \ x = 20, \frac{10}{100} \times x = 20, x = \frac{100 \times 20}{10} = 200$

2) Sophia purchased a sofa for $530.40. The sofa is regularly priced at $624. What was the percent discount Sophia received on the sofa?

Use this formula: $percent = Part \div base = 530.40 \div 624 = 0.85 = 85\%$
Therefore, the discount is: $100\% - 85\% = 15\%$

✎ *Find the selling price of each item.*

1) Original price of a computer: $500

   Tax: 6%, Selling price: $_____

2) Original price of a laptop: $350

   Tax: 8%, Selling price: $_____

3) Original price of a sofa: $800

   Tax: 7%, Selling price: $_____

4) Original price of a car: $18,500

   Tax: 8.5%, Selling price: $_____

5) Original price of a Table: $250

   Tax: 5%, Selling price: $_____

6) Original price of a house: $250,000

   Tax: 6.5% Selling price: $_____

7) Original price of a tablet: $400

   Discount: 20%, Selling price: $_____

8) Original price of a chair: $150

   Discount: 15%, Selling price: $_____

9) Original price of a book: $50

   Discount: 25%, Selling price: $_____

10) Original price of a cellphone: $500

   Discount: 10%, Selling price: $_____

Answers – Day 11

Percent of Increase and Decrease

1) 25%
2) 60%
3) 20%
4) 25%

5) 50%
6) 15%
7) 20%

Markup, Discount, and Tip

1) $530.00
2) $378.00
3) $856.00
4) $20,072.50
5) $262.50

6) $266,250
7) $320.00
8) $127.50
9) $37.50
10) $450.00

Day 12:
Exponents and Variables

Math Topics that you'll learn today:

- ✓ Multiplication Property of Exponents

- ✓ Division Property of Exponents

- ✓ Powers of Products and Quotients

Mathematics is an art of human understanding. ~ William Thurston

Multiplication Property of Exponents

Step-by-step guide:

- ✓ Exponents are shorthand for repeated multiplication of the same number by itself. For example, instead of 2×2, we can write 2^2. For $3 \times 3 \times 3 \times 3$, we can write 3^4

- ✓ In algebra, a variable is a letter used to stand for a number. The most common letters are: $x, y, z, a, b, c, m, and\ n$.

- ✓ Exponent's rules: $x^a \times x^b = x^{a+b}$, $\frac{x^a}{x^b} = x^{a-b}$

$$(x^a)^b = x^{a \times b}, \qquad (xy)^a = x^a \times y^a , (\frac{a}{b})^c = \frac{a^c}{b^c}$$

Examples:

1) Multiply. $-2x^5 \times 7x^3 =$

   Use Exponent's rules: $x^a \times x^b = x^{a+b} \rightarrow x^5 \times x^3 = x^{5+3} = x^8$

   Then: $-2x^5 \times 7x^3 = -14x^8$

2) Multiply. $(x^2y^4)^3 =$

   Use Exponent's rules: $(x^a)^b = x^{a \times b}$. Then: $(x^2y^4)^3 = x^{2 \times 3}y^{4 \times 3} = x^6y^{12}$

✎ ***Simplify and write the answer in exponential form.***

1) $x^4 \times 3x =$

2) $x \times 2x^2 =$

3) $5x^4 \times 5x^4 =$

4) $2yx^2 \times 2x =$

5) $3x^4 \times y^2x^4 =$

6) $y^2x^3 \times y^5x^2 =$

7) $4yx^3 \times 2x^2y^3 =$

8) $6x^2 \times 6x^3y^4 =$

9) $3x^4y^5 \times 7x^2y^3 =$

10) $7x^2\ ^5 \times 9xy^3 =$

11) $7xy^4 \times 4x^3y^3 =$

12) $3x^5y^3 \times 8x^2y^3 =$

Division Property of Exponents

Step-by-step guide:

✓ For division of exponents use these formulas: $\frac{x^a}{x^b} = x^{a-b}$, $x \neq 0$

$$\frac{x^a}{b} = \frac{1}{x^{b-a}} , x \neq 0, \qquad \frac{1}{x^b} = x^{-b}$$

Examples:

1) Simplify. $\frac{4x^3y}{36x^2y^3} =$

First cancel the common factor: $4 \rightarrow \frac{4x^3y}{36x^2y^3} = \frac{x^3y}{9x^2y^3}$

Use Exponent's rules: $\frac{x^a}{x^b} = x^{a-b} \rightarrow \frac{x^3}{x^2} = x^{3-b}$

Then: $\frac{4x^3y}{36x^2y^3} = \frac{xy}{9y^3} \rightarrow$ now cancel the common factor: $y \rightarrow \frac{xy}{9y^3} = \frac{x}{9y^2}$

2) Divide. $\frac{2x^{-5}}{9x^{-2}} =$

Use Exponent's rules: $\frac{x^a}{x^b} = \frac{1}{x^{b-a}} \rightarrow \frac{x^{-5}}{x^{-2}} = \frac{1}{x^{-2-(-5)}} = \frac{1}{x^{-2+5}} = \frac{1}{x^3}$

Then: $\frac{2x^{-5}}{9x^{-2}} = \frac{2}{9x^3}$

✎ *Simplify.*

1) $\frac{3^4 \times 3^7}{3^2 \times 3^8} =$

2) $\frac{5x}{10x^3} =$

3) $\frac{3x^3}{2x^5} =$

4) $\frac{12x^3}{14\ ^6} =$

5) $\frac{12x^3}{9y^8} =$

6) $\frac{25xy^4}{5x^6y^2} =$

7) $\frac{2x^4}{7x} =$

8) $\frac{16\ ^2y^8}{4x^3} =$

9) $\frac{12x^4}{15x^7y^9} =$

10) $\frac{12y\ ^4}{10yx^8} =$

11) $\frac{16x^4y}{9x^8y^2} =$

12) $\frac{5x^8}{20x^8} =$

Powers of Products and Quotients

Step-by-step guide:

✓ For any nonzero numbers a and b and any integer x, $(ab)^x = a^x \times b^x$.

Example:

1) Simplify. $(3x^5 y^4)^2 =$

Use Exponent's rules: $(x^a)^b = x^{a \times b}$

$(3x^5 y^4)^2 = (3)^2 (x^5)^2 (y^4)^2 = 9x^{5 \times 2} y^{4 \times 2} = 9x^{25} y^{16}$

2) Simplify. $\left(\frac{2x}{3x^2}\right)^2 =$

First cancel the common factor: $x \rightarrow \left(\frac{2x}{3x^2}\right)^2 = \left(\frac{2}{3x}\right)^2$

Use Exponent's rules: $\left(\frac{a}{b}\right)^c = \frac{a^c}{b^c}$

Then: $\left(\frac{2}{3x}\right)^2 = \frac{2^2}{(3x)^2} = \frac{4}{9x^2}$

✍ *Simplify.*

1) $(4x^3 x^3)^2 =$

2) $(3x^3 \times 5x)^2 =$

3) $(10x^{11} y^3)^2 =$

4) $(9x^7 {}^5)^2 =$

5) $(4x^4 y^6)^5 =$

6) $(3x \times 4y^3)^2 =$

7) $\left(\frac{5x}{x^2}\right)^2 =$

8) $\left(\frac{x^4 y^4}{x^2 y^2}\right)^3 =$

9) $\left(\frac{25x}{5x^6}\right)^2 =$

10) $\left(\frac{x^8}{x^6 y^2}\right)^2 =$

11) $\left(\frac{xy^2}{x^3 y^3}\right)^{-2} =$

12) $\left(\frac{2xy^4}{x^3}\right)^2 =$

Answers – Day 12

Multiplication Property of Exponents

1) $3x^5$
2) $2x^3$
3) $25x^8$
4) $4x^3y$
5) $3x^8y^2$
6) x^5y^7

7) $8x^5y^4$
8) $36x^5y^4$
9) $21x^6y^8$
10) $63x^3y^8$
11) $28x^4y^7$
12) $24x^7y^6$

Division Property of Exponents

1) 3
2) $\dfrac{1}{2x^2}$
3) $\dfrac{3}{2x^2}$
4) $\dfrac{6}{7}$ ^{3}
5) $\dfrac{4x^3}{3y^8}$
6) $\dfrac{5y^2}{x^5}$

7) $\dfrac{2x^3}{7}$
8) $\dfrac{4y^8}{x}$
9) $\dfrac{4}{5x^3y^9}$
10) $\dfrac{6}{5x^4}$
11) $\dfrac{16}{9x^4y}$
12) $\dfrac{1}{4}$

Powers of Products and Quotients

1) $16x^{12}$
2) $225x^8$
3) $100x^{22}y^6$
4) $81x^{14}y^{10}$
5) $1,024x^{20}y^{30}$
6) $144x^2y^6$
7) $\dfrac{25}{x^2}$

8) x^6y^6
9) $\dfrac{25}{x^{10}}$
10) $\dfrac{x^4}{y^4}$
11) x^4y^2
12) $\dfrac{4y^8}{x^4}$

Day 13:
Exponents and Roots

Math Topics that you'll learn today:

- ✓ Zero and Negative Exponents

- ✓ Negative Exponents and Negative Bases

- ✓ Scientific Notation

- ✓ Square Roots

Mathematics is an independent world created out of pure intelligence.

~ William Woods Worth

Zero and Negative Exponents

Step-by-step guide:

- ✓ A negative exponent simply means that the base is on the wrong side of the fraction line, so you need to flip the base to the other side. For instance, "x^{-2}" (pronounced as "ecks to the minus two") just means "x^2" but underneath, as in $\frac{1}{x^2}$.

Example:

1) Evaluate. $\left(\frac{4}{9}\right)^{-2} =$

Use Exponent's rules: $\frac{1}{b} = x^{-b} \rightarrow \left(\frac{4}{9}\right)^{-2} = \frac{1}{\left(\frac{4}{9}\right)^2} = \frac{1}{\frac{4^2}{9^2}}$

Now use fraction rule: $\frac{1}{\frac{b}{c}} = \frac{c}{b} \rightarrow \frac{1}{\frac{4^2}{9^2}} = \frac{9^2}{4^2} = \frac{81}{16}$

2) Evaluate. $\left(\frac{5}{6}\right)^{-3} =$

Use Exponent's rules: $\frac{1}{x^b} = x^{-b} \rightarrow \left(\frac{5}{6}\right)^{-3} = \frac{1}{\left(\frac{5}{6}\right)^3} = \frac{1}{\frac{5^3}{6^3}}$

Now use fraction rule: $\frac{1}{\frac{b}{c}} = \frac{c}{b} \rightarrow \frac{1}{\frac{5^3}{6^3}} = \frac{6^3}{5^3} = \frac{216}{125}$

✎ *Evaluate the following expressions.*

1) $2^{-3} =$

2) $3^{-3} =$

3) $7^{-3} =$

4) $6^{-3} =$

5) $8^{-3} =$

6) $9^{-2} =$

7) $10^{-3} =$

8) $10^{-9} =$

9) $\left(\frac{1}{2}\right)^{-1}$

10) $\left(\frac{1}{2}\right)^{-2} =$

11) $\left(\frac{1}{3}\right)^{-2} =$

12) $\left(\frac{2}{3}\right)^{-2} =$

Negative Exponents and Negative Bases

Step-by-step guide:

- ✓ Make the power positive. A negative exponent is the reciprocal of that number with a positive exponent.
- ✓ The parenthesis is important!
- ✓ 5^{-2} is not the same as $(-5)^{-2}$

$$(-5)^{-2} = -\frac{1}{5^2} \text{ and } (-5)^{-2} = +\frac{1}{5^2}$$

Example:

1) Simplify. $\left(\frac{3a}{2c}\right)^{-2} =$

Use Exponent's rules: $\frac{1}{x^b} = x^{-b} \rightarrow \left(\frac{3a}{2c}\right)^{-2} = \frac{1}{\left(\frac{3a}{2c}\right)^2} = \frac{1}{\frac{3^2 a^2}{2^2 c^2}}$

Now use fraction rule: $\frac{1}{\frac{b}{c}} = \frac{c}{b} \rightarrow \frac{1}{\frac{3^2 a^2}{2^2 c^2}} = \frac{2^2 c^2}{3^2 a^2}$

Then: $\frac{2^2 c^2}{3^2 a^2} = \frac{4c^2}{9a^2}$

2) Simplify. $\left(-\frac{5x}{3yz}\right)^{-3} =$

Use Exponent's rules: $\frac{1}{x^b} = x^{-b} \rightarrow \left(-\frac{5x}{3yz}\right)^{-3} = \frac{1}{\left(-\frac{5x}{3y}\right)^3} = \frac{1}{-\frac{5^3 x^3}{3^3 y^3 z^3}}$

Now use fraction rule: $\frac{1}{\frac{b}{c}} = \frac{c}{b} \rightarrow \frac{1}{-\frac{5^3 x^3}{3^3 y^3 z^3}} = -\frac{3^3 y^3 z^3}{5^3 x^3} = -\frac{27 y^3 z^3}{125 x^3}$

✍ *Simplify.*

1) $-5x^{-2} y^{-3} =$

2) $20x^{-4} y^{-1} =$

3) $14a^{-6} b^{-7} =$

4) $-12x^2 y^{-3} =$

5) $-\frac{25}{x^{-6}} =$

6) $\frac{7b}{-9c^{-4}} =$

7) $\frac{7ab}{a^{-3} b^{-1}} =$

8) $-\frac{5n^{-2}}{10p^{-3}} =$

9) $\frac{4a^{-2}}{-3c^{-2}} =$

10) $\left(\frac{3a}{2c}\right)^{-2} =$

11) $\left(-\frac{5x}{3yz}\right)^{-3} =$

12) $\frac{4ab^{-2}}{-3c^{-2}} =$

13) $\left(-\frac{x^3}{x^4}\right)^{-2} =$

Scientific Notation

Step-by-step guide:

✓ It is used to write very big or very small numbers in decimal form.
✓ In scientific notation all numbers are written in the form of:
$$m \times 10^n$$

Decimal notation	Scientific notation
5	5×10^0
– 25,000	$– 2.5 \times 10^4$
0.5	5×10^{-1}
2,122.456	$2,122456 \times 10^3$

Example:

1) Write **0.00012** in scientific notation.

First, move the decimal point to the right so that you have a number that is between 1 and 10. Then: $N = 1.2$

Second, determine how many places the decimal moved in step 1 by the power of 10. Then: $10^{-4} \rightarrow$ When the decimal moved to the right, the exponent is negative.

Then: $0.00012 = 1.2 \times 10^{-4}$

2) Write **8.3 × 10⁻⁵** in standard notation.

$10^{-5} \rightarrow$ When the decimal moved to the right, the exponent is negative.

Then: $8.3 \times 10^{-5} = 0.000083$

 *Write each number in scientific notation.*

1) $0.000325 =$ 3) $56,000,000 =$

2) $0.00023 =$ 4) $21,000 =$

Write each number in standard notation.

5) $3 \times 10^{-1} =$ 7) $1.2 \times 10^3 =$

6) $5 \times 10^{-2} =$ 8) $2 \times 10^{-4} =$

Square Roots

Step-by-step guide:

- ✓ A square root of x is a number r whose square is: $r^2 = x$

  r is a square root of x.

Example:

1) Find the square root of $\sqrt{225}$.

   First factor the number: $225 = 15^2$, Then: $\sqrt{225} = \sqrt{15^2}$

   Now use radical rule: $\sqrt[n]{a^n} = a$

   Then: $\sqrt{15^2} = 15$

2) Evaluate. $\sqrt{4} \times \sqrt{16} =$

   First factor the numbers: $4 = 2^2$ and $16 = 4^2$

   Then: $\sqrt{4} \times \sqrt{16} = \sqrt{2^2} \times \sqrt{4^2}$

   Now use radical rule: $\sqrt[n]{a^n} = a$, Then: $\sqrt{2^2} \times \sqrt{4^2} = 2 \times 4 = 8$

✎ *Evaluate.*

1) $\sqrt{4} \times \sqrt{9} =$ _____

2) $\sqrt{25} \times \sqrt{64} =$ _____

3) $\sqrt{2} \times \sqrt{8} =$ _____

4) $\sqrt{6} \times \sqrt{6} =$ _____

5) $\sqrt{5} \times \sqrt{5} =$ _____

6) $\sqrt{8} \times \sqrt{8} =$ _____

7) $\sqrt{2} + \sqrt{2} =$ _____

8) $\sqrt{8} + \sqrt{8} =$ _____

9) $4\sqrt{5} - 2\sqrt{5} =$ _____

10) $3\sqrt{3} \times 2\sqrt{3} =$ _____

11) $8\sqrt{2} \times 2\sqrt{2} =$ _____

12) $6\sqrt{3} - \sqrt{12} =$ _____

Answers – Day 13

Zero and Negative Exponents

1) $\dfrac{1}{8}$

2) $\dfrac{1}{27}$

3) $\dfrac{1}{343}$

4) $\dfrac{1}{216}$

5) $\dfrac{1}{512}$

6) $\dfrac{1}{81}$

7) $\dfrac{1}{1,000}$

8) $\dfrac{1}{1,000,000,000}$

9) 2

10) 4

11) 9

12) $\dfrac{9}{4}$

Negative Exponents and Negative Bases

1) $-\dfrac{5}{x^2}^3$

2) $\dfrac{20}{x^4 y}$

3) $\dfrac{14}{a^6 b^7}$

4) $-\dfrac{12x^2}{y^3}$

5) $-25x^6$

6) $-\dfrac{7bc^4}{9}$

7) $7a^4 b^2$

8) $-\dfrac{p^3}{2n^2}$

9) $-\dfrac{4ac^2}{3b^2}$

10) $\dfrac{4c^2}{9a^2}$

11) $-\dfrac{27\,^3 z^3}{125 x^3}$

12) $-\dfrac{4a\,^2}{3b^2}$

13) ^{2}

Scientific Notation

1) 3.25×10^{-4}

2) 2.3×10^{-4}

3) 5.6×10^{7}

4) 2.1×10^{4}

5) 0.3

6) 0.05

7) $1,200$

8) 0.0002

Square Roots

1) 6

2) 40

3) 4

4) 6

5) 5

6) 8

7) $2\sqrt{2}$

8) $2\sqrt{8}$

9) $2\sqrt{5}$

10) 18

11) 32

12) $4\sqrt{3}$

Day 14:
Expressions and
Variables

Math Topics that you'll learn today:

✓ Simplifying Variable Expressions

✓ Simplifying Polynomial Expressions

✓ Translate Phrases into an Algebraic Statement

Mathematics is, as it were, a sensuous logic, and relates to philosophy as do the arts, music, and plastic art to poetry. – K.

Shegel

Simplifying Variable Expressions

Step-by-step guide:

- ✓ In algebra, a variable is a letter used to stand for a number. The most common letters are: $x, y, z, a, b, c, m, and\ n$.
- ✓ algebraic expression is an expression contains integers, variables, and the math operations such as addition, subtraction, multiplication, division, etc.
- ✓ In an expression, we can combine "like" terms. (values with same variable and same power)

Examples:

1) Simplify this expression. $(10x + 2x + 3) =$?
   Combine like terms. Then: $(10x + 2x + 3) = 12x + 3$ (remember you cannot combine variables and numbers.

2) Simplify this expression. $12 - 3x^2 + 9x + 5x^2 =$?
   Combine "like" terms: $-3x^2 + 5x^2 = 2x^2$

   Then: $12 - 3x^2 + 9x + 5x^2 = 12 + 2x^2 + 9x$. Write in standard form (biggest powers first): $2x^2 + 9x + 12$

✐ *Simplify each expression.*

1) $(2x + x + 3 + 24) =$

2) $(-28x - 20x + 24) =$

3) $7x + 3 - 3x =$

4) $-2 - x^2 - 6x^2 =$

5) $3 + 10x^2 + 2 =$

6) $8x^2 + 6x + 7x^2 =$

7) $5x^2 - 12x^2 + 8x =$

8) $2x^2 - 2x - x =$

9) $4x + (12 - 30x) =$

10) $10x + (80x - 48) =$

11) $(-18x - 54) - 5 =$

12) $2x^2 + (-8x) =$

Simplifying Polynomial Expressions

Step-by-step guide:

✓ In mathematics, a polynomial is an expression consisting of variables and coefficients that involves only the operations of addition, subtraction, multiplication, and non-negative integer exponents of variables.

$$P(x) = a_n x^n + a_{n-1} x^{n-1} + \ldots + a_2 x^2 + a_1 x + a_0$$

Examples:

1) Simplify this Polynomial Expressions. $4x^2 - 5x^3 + 15x^4 - 12x^3 =$
   Combine "like" terms: $-5x^3 - 12x^3 = -17x^3$
   Then: $4x^2 - 5x^3 + 15x^4 - 12x^3 = 4x^2 - 17x^3 + 15x^4$
   Then write in standard form: $4x^2 - 17x^3 + 15x^4 = 15x^4 - 17x^3 + 4x^2$

2) Simplify this expression. $(2x^2 - x^4) - (4x^4 - x^2) =$
   First use distributive property: $\rightarrow$ multiply $(-)$ into $(4x^4 - x^2)$
   $(2x^2 - x^4) - (4x^4 - x^2) = 2x^2 - x^4 - 4x^4 + x^2$
   Then combine "like" terms: $2x^2 - x^4 - 4x^4 + x^2 = 3x^2 - 5x^4$
   And write in standard form: $3x^2 - 5x^4 = -5x^4 + 3x^2$

✍ *Simplify each polynomial.*

1) $(2x^3 + 5x^2) - (12x + 2x^2) =$ _____

2) $(2x^5 + 2x^3) - (7x^3 + 6x^2) =$ _____

3) $(12x^4 + 4x^2) - (2x^2 - 6x^4) =$ _____

4) $14x - 3x^2 - 2(6x^2 + 6x^3) =$ _____

5) $(5x^3 - 3) + 5(2x^2 - 3x^3) =$ _____

6) $(4x^3 - 2x) - 2(4x^3 - 2x^4) =$ _____

7) $2(4x - 3x^3) - 3(3x^3 + 4x^2) =$ _____

8) $(2x^2 - 2x) - (2x^3 + 5x^2) =$ _____

Translate Phrases into an Algebraic Statement

Step-by-step guide:

Translating key words and phrases into algebraic expressions:

- ✓ Addition: plus, more than, the sum of, etc.
- ✓ Subtraction: minus, less than, decreased, etc.
- ✓ Multiplication: times, product, multiplied, etc.
- ✓ Division: quotient, divided, ratio, etc.

Examples:

Write an algebraic expression for each phrase.

1) Eight more than a number is 20.

More than mean plus a number $= x$

Then: $8 + x = 20$

2) 5 times the sum of 8 and x.

Sum of 8 and x: $8 + x$. Times means multiplication. Then: $5 \times (8 + x)$

✍ *Write an algebraic expression for each phrase.*

1) 4 multiplied by x. _____

2) Subtract 8 from y. _____

3) 6 divided by x. _____

4) 12 decreased by y. _____

5) Add y to 9. _____

6) The square of 5. _____

7) x raised to the fourth power. _____

8) The sum of nine and a number. _____

9) The difference between sixty–four and y. _____

10) The quotient of twelve and a number. _____

11) The quotient of the square of x and 7. _____

12) The difference between x and 8 is 22. _____

Answers – Day 14

Simplifying Variable Expressions

1) $3x + 27$

2) $-48x + 24$

3) $4x + 3$

4) $-7x^2 - 2$

5) $10x^2 + 5$

6) $15x^2 + 6x$

7) $-7x^2 + 8x$

8) $2x^2 - 3x$

9) $-26x + 12$

10) $90x - 48$

11) $-18x - 59$

12) $2x^2 - 8x$

Simplifying Polynomial Expressions

1) $2x^3 + 3x^2 - 12x$

2) $2x^5 - 5x^3 - 6x^2$

3) $18x^4 + 2x^2$

4) $-12x^3 - 15x^2 + 14x$

5) $-10x^3 + 10x^2 - 3$

6) $4x^4 - 4x^3 - 2$

7) $-15x^3 - 12x^2 + 8x$

8) $-2x^3 - 3x^2 - 2x$

Translate Phrases into an Algebraic Statement

1) $4x$

2) $y - 8$

3) $\frac{6}{x}$

4) $12 - y$

5) $y + 9$

6) 5^2

7) x^4

8) $9 + x$

9) $64 - y$

10) $\frac{12}{x}$

11) $\frac{x^2}{7}$

12) $x - 8 = 22$

Day 15:
Evaluating Variables

Math Topics that you'll learn today:

- ✓ The Distributive Property

- ✓ Evaluating One Variable

- ✓ Evaluating Two Variables

- ✓ Combining like Terms

Mathematics is on the artistic side a creation of new rhythms, orders, designs, harmonies, and on the knowledge side, is a systematic study of various rhythms, orders. - William L. Schaaf

The Distributive Property

Step-by-step guide:

✓ Distributive Property:
$$a(b + c) = ab + ac$$

Examples:

1) Simply. $(5x - 3)(-5) =$

Use Distributive Property formula: $a(b + c) = ab + ac$
$(5x - 3)(-5) = -25x + 15$

2) Simply$(-8)(2x - 8) =$

Use Distributive Property formula: $a(b + c) = ab + ac$
$(-8)(2x - 8) = -16x + 64$

✍ *Use the distributive property to simply each expression.*

1) $2(2 + 3x) =$

2) $3(5 + 5x) =$

3) $4(3x - 8) =$

4) $(6x - 2)(-2) =$

5) $(-3)(x + 2) =$

6) $(2 + 2x)5 =$

7) $(-4)(4 - 2x) =$

8) $-(-2 - 5x) =$

9) $(-6x + 2)(-1) =$

10) $(-5)(x - 2) =$

11) $-(7 - 3x) =$

12) $8(8 + 2x) =$

Evaluating One Variable

Step-by-step guide:

- ✓ To evaluate one variable expression, find the variable and substitute a number for that variable.
- ✓ Perform the arithmetic operations.

Examples:

1) Solve this expression. $12 - 2x$, $x = -1$

First substitute -1 for x, then:

$12 - 2x = 12 - 2(-1) = 12 + 2 = 14$

2) Solve this expression. $-8 + 5x$, $x = 3$

First substitute 3 for x, then:

$-8 + 5x = -8 + 5(3) = -8 + 15 = 7$

✎ *Evaluate each expression using the value given.*

1) $5 + x$, $x = 2$

2) $x - 2, x = 4$

3) $8x + 1, x = 9$

4) $x - 12, x = -1$

5) $9 - x$, $x = 3$

6) $x + 2, x = 5$

7) $3x + 7, x = 6$

8) $x + (-5), x = -2$

9) $3x + 6, x = 4$

10) $4x + 6, x = -1$

11) $10 + 2x - 6, x = 3$

12) $10 - 3x, x = 8$

Evaluating Two Variables

Step-by-step guide:

✓ To evaluate an algebraic expression, substitute a number for each variable and perform the arithmetic operations.

Examples:

1) Solve this expression. $-3x + 5y$, $x = 2, y = -1$

First substitute 2 for x, and -1 for y, then:

$-3x + 5y = -3(2) + 5(-1) = -6 - 5 = -11$

2) Solve this expression. $2(a - 2b), a = -1, b = 3$

First substitute -1 for a, and 3 for b, then:

$2(a - 2b) = 2a - 4b = 2(-1) - 4(3) = -2 - 12 = -14$

✎ *Evaluate each expression using the values given.*

1) $2x + 4y$,

   $x = 3, y = 2$

2) $8x + 5y$,

   $x = 1, y = 5$

3) $-2a + 4b$,

   $a = 6, b = 3$

4) $4x + 7 - 2y$,

   $x = 7, y = 6$

5) $5z + 12 - 4k$,

   $z = 5, k = 2$

6) $2(-x - 2y)$,

   $x = 6, y = 9$

7) $18a + 2b$,

   $a = 2, b = 8$

8) $4x \div 3y$,

   $x = 3, y = 2$

9) $2x + 15 + 4y$,

   $x = -2, y = 4$

10) $4a - (15 - b)$,

   $a = 4, b = 6$

11) $5z + 19 + 8k$,

   $z = -5, k = 4$

12) $xy + 12 + 5x$,

   $x = 7, y = 2$

Combining like Terms

Step-by-step guide:

- ✓ Terms are separated by "+" and "–" signs.
- ✓ Like terms are terms with same variables and same powers.
- ✓ Be sure to use the "+" or "–" that is in front of the coefficient.

Examples:

1) Simplify this expression. $(-5)(8x - 6) =$

Use Distributive Property formula: $a(b + c) = a + ac$
$(-5)(8x - 6) = -40x + 30$

2) Simplify this expression. $(-3)(2x - 2) + 6 =$

First use Distributive Property formula: $a(b + c) = ab + ac$
$(-3)(2x - 2) + 6 = -6x + 6 + 6$

And Combining like Terms:

$-6x + 6 + 6 = -6x + 12$

✍ *Simplify each expression.*

1) $2x + x + 2 =$

2) $2(5x - 3) =$

3) $7x - 2x + 8 =$

4) $(-4)(3x - 5) =$

5) $9x - 7x - 5 =$

6) $16x - 5 + 8x =$

7) $5 - (5x + 6) =$

8) $-12x + 7 - 10x =$

9) $7x - 11 - 2x + 2 =$

10) $12x + 4x - 21 =$

11) $5 + 2x - 8 =$

12) $(-2x + 6)^2 =$

Answers – Day 15

The Distributive Property

1) $6x + 4$
2) $15x + 15$
3) $12x - 32$
4) $-12x + 4$
5) $-3x - 6$
6) $10x + 10$

7) $8x - 16$
8) $5x + 2$
9) $6x - 2$
10) $-5x + 10$
11) $3x - 7$
12) $16x + 64$

Evaluating One Variable

1) 7
2) 2
3) 73
4) -13
5) 6
6) 7

7) 25
8) -7
9) 18
10) 2
11) 10
12) -14

Evaluating Two Variables

1) 14
2) 33
3) 0
4) 23
5) 29
6) -48

7) 52
8) 2
9) 27
10) 7
11) 26
12) 61

Combining like Terms

1) $3x + 2$
2) $10x - 6$
3) $5x + 8$
4) $-12x + 20$
5) $2x - 5$
6) $24x - 5$

7) $-5x - 1$
8) $-22x + 7$
9) $5x - 9$
10) $16x - 21$
11) $2x - 3$
12) $4x^2 - 24x + 36$

Day 16:
Equations and Inequalities

Math Topics that you'll learn today:

- ✓ One–Step Equations

- ✓ Multi–Step Equations

- ✓ Graphing Single–Variable Inequalities

"Life is a math equation. In order to gain the most, you have to know how to convert negatives into positives."

– Anonymous

One–Step Equations

Step-by-step guide:

✓ The values of two expressions on both sides of an equation are equal. $ax + b = c$
✓ You only need to perform one Math operation in order to solve the one-step equations.
✓ To solve one-step equation, find the inverse (opposite) operation is being performed.
✓ The inverse operations are:
  - Addition and subtraction
  - Multiplication and division

Examples:

1) Solve this equation. $x + 24 = 0, x = ?$
   Here, the operation is addition and its inverse operation is subtraction. To solve this equation, subtract 24 from both sides of the equation: $x + 24 - 24 = 0 - 24$
   Then simplify: $x + 24 - 24 = 0 - 24 \rightarrow x = -24$

2) Solve this equation. $3x = 15, x = ?$
   Here, the operation is multiplication (variable x is multiplied by 3) and its inverse operation is division. To solve this equation, divide both sides of equation by 3:
   $$3x = 15 \rightarrow 3x \div 3 = 15 \div 3 \rightarrow x = 5$$

✎ *Solve each equation.*

1) $16 = -4 + x, x =$ ____

2) $x - 4 = -25, x =$ ____

3) $x + 12 = -9, x =$ ____

4) $14 = 18 - x, x =$ ____

5) $2 + x = -14, x =$ ____

6) $x - 5 = 15, x =$ ____

7) $25 = x - 5, x =$ ____

8) $x - 3 = -12, x =$ ____

9) $x - 12 = 12, x =$ ____

10) $x - 12 = -25, x =$ ____

11) $x - 13 = 32, x =$ ____

12) $-55 = x - 18, x =$ ____

Multi–Step Equations

Step-by-step guide:

- ✓ Combine "like" terms on one side.
- ✓ Bring variables to one side by adding or subtracting.
- ✓ Simplify using the inverse of addition or subtraction.
- ✓ Simplify further by using the inverse of multiplication or division.

Examples:

1) Solve this equation. $-(2 - x) = 5$

First use Distributive Property: $-(2 - x) = -2 + x$

Now solve by adding 2 to both sides of the equation. $-2 + x = 5 \rightarrow -2 + x + 2 = 5 + 2$

Now simplify: $-2 + x + 2 = 5 + 2 \rightarrow x = 7$

2) Solve this equation. $4x + 10 = 25 - x$

First bring variables to one side by adding x to both sides.

$4x + 10 + x = 25 - x + x \rightarrow 5x + 10 = 25$. Now, subtract 10 from both sides:

$5x + 10 - 10 = 25 - 10 \rightarrow 5x = 15$

Now, divide both sides by 5: $5x = 15 \rightarrow 5x \div 5 = \frac{15}{5} \rightarrow x = 3$

✍ *Solve each equation.*

1) $-3(2 + x) = 3$

2) $-2(4 + x) = 4$

3) $20 = -(x - 8)$

4) $2(2 - 2x) = 20$

5) $-12 = -(2x + 8)$

6) $5(2 + x) = 5$

7) $2(x - 14) = 4$

8) $-28 = 2x + 12x$

9) $3x + 15 = -x - 5$

10) $2(3 + 2x) = -18$

11) $12 - 2x = -8 - x$

12) $10 - 3x = 14 + x$

Graphing Single–Variable Inequalities

Step-by-step guide:

- ✓ Inequality is similar to equations and uses symbols for "less than" (<) and "greater than" (>).
- ✓ To solve inequalities, we need to isolate the variable. (like in equations)
- ✓ To graph an inequality, find the value of the inequality on the number line.
- ✓ For less than or greater than draw open circle on the value of the variable.
- ✓ If there is an equal sign too, then use filled circle.
- ✓ Draw a line to the right or to the left for greater or less than.

Examples:

1) Draw a graph for $x > 2$

Since, the variable is greater than 2, then we need to find 2 and draw an open circle above it. Then, draw a line to the right.

Graph this inequality. $x < 5$

✎ *Draw a graph for each inequality.*

1) $x > -1$

2) $x < 3$

3) $x < -5$

4) $x > -2$

5) $x < 0$

Answers – Day 16

One–Step Equations

1) 20
2) −21
3) −21
4) 4
5) −16
6) 20

7) 30
8) −9
9) 24
10) −13
11) 45
12) −37

Multi–Step Equations

1) −3
2) −6
3) −12
4) −4
5) 2
6) −1

7) 16
8) −2
9) −5
10) −6
11) 20
12) −1

Graphing Single–Variable Inequalities

1)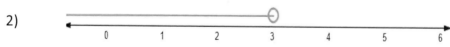

2)

3)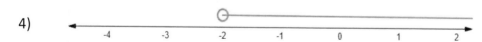

4)

5)

Day 17:
Solving Inequalities

Math Topics that you'll learn today:

- ✓ One–Step Inequalities

- ✓ Multi–Step Inequalities

"Go down deep enough into anything and you will find mathematics." - Dean Schlicter

One–Step Inequalities

Step-by-step guide:

✓ Similar to equations, first isolate the variable by using inverse operation.
✓ For dividing or multiplying both sides by negative numbers, flip the direction of the inequality sign.

Examples:

1) Solve and graph the inequality. $x + 2 \geq 3$.

Subtract 2 from both sides. $x + 2 \geq 3 \rightarrow x + 2 - 2 \geq 3 - 2$, then: $x \geq 1$

2) Solve this inequality. $x - 1 \leq 2$

Add 1 to both sides. $x - 1 \leq 2 \rightarrow x - 1 + 1 \leq 2 + 1$, then: $x \leq 3$

 Solve each inequality and graph it.

1) $2x \geq 12$

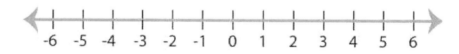

2) $4 + x \leq 5$

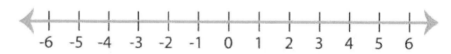

3) $x + 3 \leq -3$

4) $4x \geq 16$

5) $9x \leq 18$

Multi–Step Inequalities

Step-by-step guide:

- ✓ Isolate the variable.
- ✓ Simplify using the inverse of addition or subtraction.
- ✓ Simplify further by using the inverse of multiplication or division.

Examples:

1) Solve this inequality. $2x - 2 \leq 6$

   First add 2 to both sides: $2x - 2 + 2 \leq 6 + 2 \rightarrow 2x \leq 8$

   Now, divide both sides by 2: $2x \leq 8 \rightarrow x \leq 4$

2) Solve this inequality. $2x - 4 \leq 8$

   First add 4 to both sides: $2x - 4 + 4 \leq 8 + 4$

   Then simplify: $2x - 4 + 4 \leq 8 + 4 \rightarrow 2x \leq 12$

   Now divide both sides by 2: $\frac{2x}{2} \leq \frac{12}{2} \rightarrow x \leq 6$

✍ *Solve each inequality.*

1) $2x - 8 \leq 6$

2) $8x - 2 \leq 14$

3) $-5 + 3x \leq 10$

4) $2(x - 3) \leq 6$

5) $7x - 5 \leq 9$

6) $4x - 21 < 19$

7) $2x - 3 < 21$

8) $17 - 3x \geq -13$

9) $9 + 4x < 21$

10) $3 + 2x \geq 19$

11) $6 + 2x < 32$

12) $4x - 1 < 7$

Answers – Day 17

One–Step Inequalities

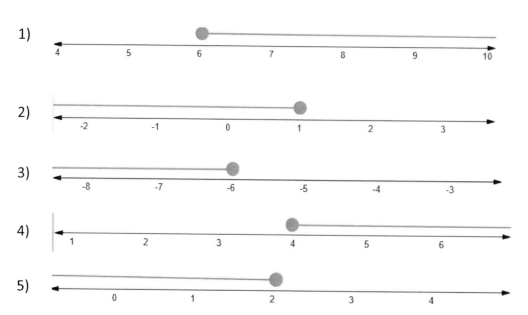

Multi–Step inequalities

1) $x \leq 7$
2) $x \leq 2$
3) $x \leq 5$
4) $x \leq 6$
5) $x \leq 2$
6) $x < 10$
7) $x < 12$
8) $x \geq 10$ $x \leq 10$
9) $x < 3$
10) $x \geq 8$
11) $x < 13$
12) $x < 2$

Day 18:
Lines and Slope

Math Topics that you'll learn today:

- ✓ Finding Slope

- ✓ Graphing Lines Using Slope–Intercept Form

- ✓ Graphing Lines Using Standard Form

"Nature is written in mathematical language." - Galileo Galilei

Finding Slope

Step-by-step guide:

- ✓ The slope of a line represents the direction of a line on the coordinate plane.
- ✓ A coordinate plane contains two perpendicular number lines. The horizontal line is x and the vertical line is y. The point at which the two axes intersect is called the origin. An ordered pair (x, y) shows the location of a point.
- ✓ A line on coordinate plane can be drawn by connecting two points.
- ✓ To find the slope of a line, we need two points.
- ✓ The slope of a line with two points A (x_1, y_1) and B (x_2, y_2) can be found by using this formula: $\dfrac{y_2 - y_1}{x_2 - x_1} = \dfrac{rise}{run}$

Examples:

1) Find the slope of the line through these two points: $(2, -10)$ *and* $(3, 6)$.

    Slope $= \dfrac{y_2 - y_1}{x_2 - x_1}$. Let (x_1, y_1) be $(2, -10)$ and (x_2, y_2) be $(3, 6)$. Then: slope $= \dfrac{y_2 - y_1}{x_2 - x_1} = \dfrac{6 - (-10)}{3 - 2} = \dfrac{6 + 10}{1} = \dfrac{16}{1} = 16$

2) Find the slope of the line containing two points $(8, 3)$ and $(-4, 9)$.

    Slope $= \dfrac{y_2 - y_1}{x_2 - x_1} \rightarrow (x_1, y_1) = (8, 3)$ and $(x_2, y_2) = (-4, 9)$. Then: slope $= \dfrac{y_2 - y_1}{x_2 - x_1} = \dfrac{9 - 3}{-4 - 8} = \dfrac{6}{-12} = \dfrac{1}{-2} = -\dfrac{1}{2}$

✍ *Find the slope of the line through each pair of points.*

1) $(1, 1), (2, 3)$

2) $(-1, 2), (0, 3)$

3) $(3, -1), (2, 3)$

4) $(-2, -1), (0, 5)$

5) $(5, 1), (2, 4)$

6) $(-3, 1), (-2, 4)$

7) $(6, 2), (7, 4)$

8) $(6, -5), (3, 4)$

9) $(12, -9), (11, -8)$

10) $(7, 4), (5, -2)$

11) $(1, 1), (3, 5)$

12) $(7, -12), (5, 10)$

Graphing Lines Using Slope–Intercept Form

Step-by-step guide:

✓ Slope-intercept form of a line: given the slope m and the y-intercept (the intersection of the line and y-axis) b, then the equation of the line is:

$$y = mx + b$$

Example: *Sketch the graph of* $y = 8x - 3$.

To graph this line, we need to find two points. When x is zero the value of y is -3. And when y is zero the value of x is 3/8. $x = 0 \rightarrow y = 8(0) - 3 = -3, y = 0 \rightarrow 0 = 8x - 3 \rightarrow x = \frac{3}{8}$

Now, we have two points: $(0, -3)$ and $(\frac{3}{8}, 0)$. Find the points and graph the line. Remember that the slope of the line is 8.

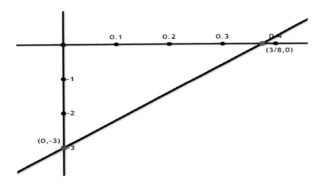

✍ ***Sketch the graph of each line.***

1) $y = \frac{1}{2}x - 4$

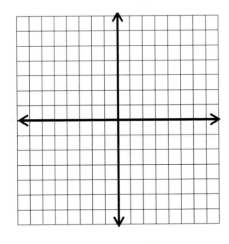

2) $y = 2x$

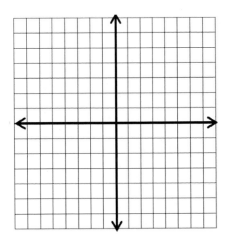

Graphing Lines Using Standard Form

Step-by-step guide:

- ✓ Find the $x-$intercept of the line by putting zero for y.
- ✓ Find the $y-$intercept of the line by putting zero for the x.
- ✓ Connect these two points.

Example:

Sketch the graph of $x - y = -5$.

First isolate y for x: $x - y = -5 \rightarrow y = x + 5$

Find the x–intercept of the line by putting zero for y.

$y = x + 5 \rightarrow x + 5 = 0 \rightarrow x = -5$

Find the y–intercept of the line by putting zero for the x.

$y = 0 + 5 \rightarrow y = 5$

Then: x–intercept: $(-5,0)$ and y–intercept: $(0,5)$

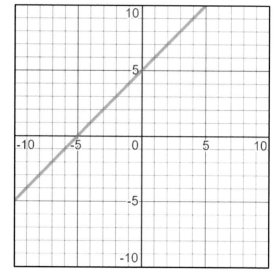

✎ *Sketch the graph of each line.*

1) $y = 3x - 2$ 2) $y = -x + 1$ 3) $x + y = 4$

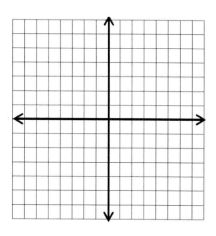 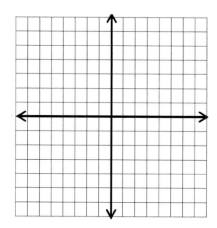 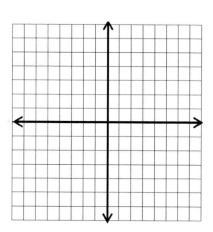

Answers – Day 18

Finding Slope

1) 2
2) 1
3) −4
4) 3
5) −1
6) 3

7) 2
8) −3
9) −1
10) 3
11) 2
12) −11

Graphing Lines Using Slope–Intercept Form

1)

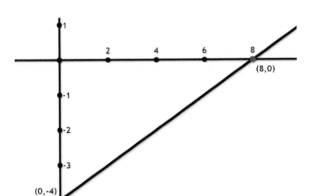

2)

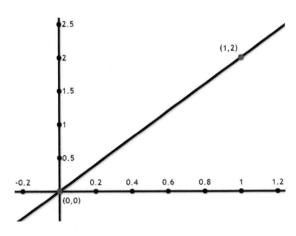

Graphing Lines Using Standard Form

1) $y = 3x - 2$

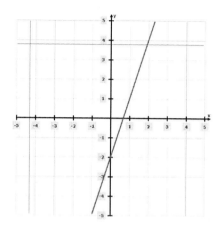

2) $y = -x + 1$

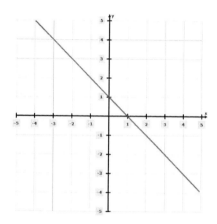

3) $x + y = 4$

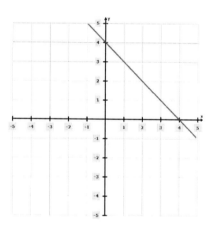

Day 19:
Linear Equations and Inequalities

Math Topics that you'll learn today:

- ✓ Writing Linear Equations

- ✓ Graphing Linear Inequalities

- ✓ Finding Midpoint

- ✓ Finding Distance of Two Points

Mathematics is a game played according to certain rules with meaningless marks on paper. ~ David Hilbert

Writing Linear Equations

Step-by-step guide:

- ✓ The equation of a line: $y = mx + b$
- ✓ Identify the slope.
- ✓ Find the y-intercept. This can be done by substituting the slope and the coordinates of a point (x, y) on the line.

Example:

1) What is the equation of the line that passes through $(2, -2)$ and has a slope of 7?

The general slope-intercept form of the equation of a line is $y = mx + b$, where m is the slope and b is the y-intercept.

By substitution of the given point and given slope, we have: $-2 = (2)(7) + b$

So, $b = -2 - 14 = -16$, and the required equation is $y = 2x - 16$.

2) Write the equation of the line through $(2, 1)$ and $(-1, 4)$.

$Slop = \frac{y_2 - y_1}{x_2 - x_1} = \frac{4 - 1}{-1 - 2} = \frac{3}{-3} = -1 \rightarrow m = -1$

To find the value of b, you can use either points. The answer will be the same: $y = -x + b$

$(2, 1) \rightarrow 1 = -2 + b \rightarrow b = 3$

$(-1, 4) \rightarrow 4 = -(-1) + b \rightarrow b = 3$

The equation of the line is: $y = -x + 3$

✐ ***Write the equation of the line through the given points.***

1) through: $(1, -2), (2, 3)$

2) through: $(-2, 1), (1, 4)$

3) through: $(-2, 1), (0, 5)$

4) through: $(5, 4), (2, 1)$

5) through: $(-4, 9), (3, 2)$

6) through: $(8, 3), (7, 2)$

7) through: $(7, -2), (5, 2)$

8) through: $(-3, 9), (5, -7)$

Graphing Linear Inequalities

Step-by-step guide:

- ✓ First, graph the "equals" line.
- ✓ Choose a testing point. (it can be any point on both sides of the line.)
- ✓ Put the value of (x, y) of that point in the inequality. If that works, that part of the line is the solution. If the values don't work, then the other part of the line is the solution.

Example:

Sketch the graph of $y < 2x - 3$. First, graph the line:

$y = 2x - 3$. The slope is 2 and y-intercept is -3. Then, choose a testing point. The easiest point to test is the origin: $(0, 0)$

$$(0,0) \rightarrow y < 2x - 3 \rightarrow 0 < 2(0) - 3 \rightarrow 0 < -3$$

0 is not less than -3. So, the other part of the line (on the right side) is the solution.

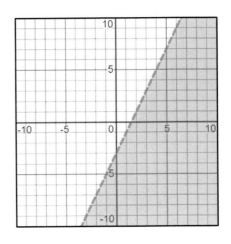

✍ *Sketch the graph of each linear inequality.*

1) $y > 3x - 1$ 2) $y < -x + 4$ 3) $y \leq -5x + 8$

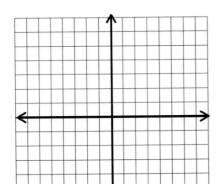 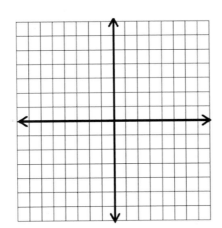 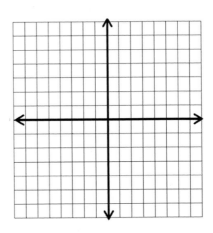

Finding Midpoint

Step-by-step guide:

- ✓ The middle of a line segment is its midpoint.
- ✓ The Midpoint of two endpoints A (x_1, y_1) and B (x_2, y_2) can be found using this formula: $M\left(\frac{x_1+x_2}{2}, \frac{y_1+y_2}{2}\right)$

Example:

1) Find the midpoint of the line segment with the given endpoints. $(4, -5), (0, 9)$

Midpoint $= \left(\frac{x_1+x_2}{2}, \frac{y_1+y_2}{2}\right) \rightarrow (x_1, y_1) = (4, -5)$ and $(x_2, y_2) = (0, 9)$

Midpoint $= \left(\frac{4+0}{2}, \frac{-5+9}{2}\right) \rightarrow \left(\frac{4}{2}, \frac{4}{2}\right) \rightarrow M(2, 2)$

2) Find the midpoint of the line segment with the given endpoints. $(6, 7), (4, -5)$

Midpoint $= \left(\frac{x_1+x_2}{2}, \frac{y_1+y_2}{2}\right) \rightarrow (x_1, y_1) = (6, 7)$ and $(x_2, y_2) = (4, -5)$

Midpoint $= \left(\frac{6+4}{2}, \frac{7-5}{2}\right) \rightarrow \left(\frac{10}{2}, \frac{2}{2}\right) \rightarrow (5, 1)$

✎ *Find the midpoint of the line segment with the given endpoints.*

1) $(-2, -2), (0, 2)$

2) $(5, 1), (-2, 4)$

3) $(4, -1), (0, 3)$

4) $(-3, 5), (-1, 3)$

5) $(3, -2), (7, -6)$

6) $(-4, -3), (2, -7)$

7) $(5, 0), (-5, 8)$

8) $(-6, 4), (-2, 0)$

9) $(-3, 4), (9, -6)$

10) $(2, 8), (6, -2)$

11) $(4, 7), (-6, 5)$

12) $(9, 3), (-1, -7)$

Finding Distance of Two Points

Step-by-step guide:

✓ Distance of two points A (x_1, y_1) and B (x_2, y_2): $d = \sqrt{(x_1 - x_2)^2 + (y_1 - y_2)^2}$

Example:

1) Find the distance between of $(0, 8), (-4, 5)$.

Use distance of two points formula: $d = \sqrt{(x_1 - x_2)^2 + (y_1 - y_2)^2}$

$(x_1, y_1) = (0, 8)$ and $(x_2, y_2) = (-4, 5)$. Then: $d = \sqrt{(x_1 - x_2)^2 + (y_1 - y_2)^2} \rightarrow$

$d = \sqrt{(0 - (-4))^2 + (8 - 5)^2} = \sqrt{(4)^2 + (3)^2} = \sqrt{16 + 9} = \sqrt{25} = 5 \rightarrow d = 5$

2) Find the distance of two points $(4, 2)$ and $(-5, -10)$.

Use distance of two points formula: $d = \sqrt{(x_1 - x_2)^2 + (y_1 - y_2)^2}$

$(x_1, y_1) = (4, 2)$, and $(x_2, y_2) = (-5, -10)$

Then: $d = \sqrt{(x_1 - x_2)^2 + (y_1 - y_2)^2} \rightarrow d = \sqrt{(4 - (-5))^2 + (2 - (-10))^2} =$

$\sqrt{(9)^2 + (12)^2} = \sqrt{81 + 144} = \sqrt{225} = 15$. Then: $d = 15$

✍ *Find the distance between each pair of points.*

1) $(2, 1), (-1, -3)$

2) $(-2, -1), (2, 2)$

3) $(-1, 0), (5, 8)$

4) $(-4, -1), (1, 11)$

5) $(3, -2), (-6, -14)$

6) $(-6, 0), (-2, 3)$

7) $(3, 2), (11, 17)$

8) $(-6, -10), (6, -1)$

9) $(5, 9), (-11, -3)$

10) $(9, -3), (3, -11)$

11) $(2, 0), (12, 24)$

12) $(8, 4), (3, -8)$

Answers – Day 19

Writing Linear Equations

1) $y = 5x - 7$
2) $y = x + 3$
3) $y = 2x + 5$
4) $y = x - 1$

5) $y = -x + 5$
6) $y = x - 5$
7) $y = -2x + 12$
8) $y = -2x + 3$

Graphing Linear Inequalities

1) $y > 3x - 1$

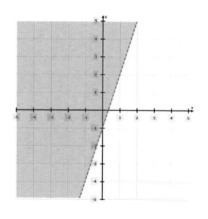

2) $y < -x + 4$

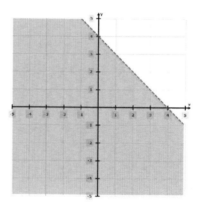

3) $y \leq -5x + 8$

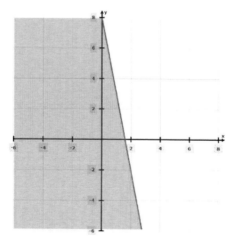

Finding Midpoint

1) $(-1, 0)$	5) $(5, -4)$	9) $(3, -1)$
2) $(1.5, 2.5)$	6) $(-1, -5)$	10) $(4, 3)$
3) $(2, 1)$	7) $(0, 4)$	11) $(-1, 6)$
4) $(-2, 4)$	8) $(-4, 2)$	12) $(4, -2)$

Finding Distance of Two Points

1) 5	5) 15	9) 20
2) 5	6) 5	10) 10
3) 10	7) 17	11) 26
4) 13	8) 15	12) 13

Day 20:
Polynomials

Math Topics that you'll learn today:

- ✓ Writing Polynomials in Standard Form

- ✓ Simplifying Polynomials

- ✓ Adding and Subtracting Polynomials

Mathematics is the supreme judge: from its decisions there is no appeal. - Tobias Dantzig

Writing Polynomials in Standard Form

Step-by-step guide:

- ✓ A polynomial function $f(x)$ of degree n is of the form
$$f(x) = a_n x^n + a_{n-1} x_{n-1} + \cdots + a_1 x + a_0$$
- ✓ The first term is the one with the biggest power!

Example:

1) Write this polynomial in standard form. $-12 + 3x^2 - 6x^4 =$

The first term is the one with the biggest power: $-12 + 3x^2 - 6x^4 = -6x^4 + 3x^2 - 12$

2) Write this polynomial in standard form. $5x^2 - 9x^5 + 8x^3 - 11 =$

The first term is the one with the biggest power: $5x^2 - 9x^5 + 8x^3 - 11 =$
$-9x^5 + 8x^3 + 5x^2 - 11$

✎ *Write each polynomial in standard form.*

1) $9x - 7x =$

2) $-3 + 16x - 16x =$

3) $3x^2 - 5x^3 =$

4) $3 + 4x^3 - 3 =$

5) $2x^2 + 1x - 6x^3 =$

6) $-x^2 + 2x^3 =$

7) $2x + 4x^3 - 2x^2 =$

8) $-2x^2 + 4x - 6x^3 =$

9) $2x^2 + 2 - 5x =$

10) $12 - 7x + 9x^4 =$

11) $5x^2 + 13x - 2x^3 =$

12) $10 + 6x^2 - x^3 =$

Simplifying Polynomials

Step-by-step guide:

✓ Find "like" terms. (they have same variables with same power).

✓ Use "FOIL". (First–Out–In–Last) for binomials:

$$(x + a)(x + b) = x^2 + (b + a)x + ab$$

✓ Add or Subtract "like" terms using order of operation.

Example:

1) Simplify this expression. $4x(6x - 3) =$

   Use Distributive Property: $4x(6x - 3) = 24x^2 - 12x$

2) Simplify this expression. $(6x - 2)(2x - 3) =$

   First apply FOIL method: $(a + b)(c + d) = ac + ad + bc + bd$

   $(6x - 2)(2x - 3) = 12x^2 - 18x - 4x + 6$

   Now combine like terms: $12x^2 - 18x - 4x + 6 = 12x^2 - 22x + 6$

✎ *Simplify each expression.*

1) $5(2x - 10) =$

2) $2x(4x - 2) =$

3) $4x(5x - 3) =$

4) $3x(7x + 3) =$

5) $4x(8x - 4) =$

6) $5x(5x + 4) =$

7) $(2x - 3)(x - 4) =$

8) $(x - 5)(3x + 4) =$

9) $(x - 5)(x - 3) =$

10) $(3x + 8)(3x - 8) =$

11) $(3x - 8)(3x - 4) =$

12) $3x^2 + 3x^2 - 2x^3 =$

Adding and Subtracting Polynomials

Step-by-step guide:

- ✓ Adding polynomials is just a matter of combining like terms, with some order of operations considerations thrown in.
- ✓ Be careful with the minus signs, and don't confuse addition and multiplication!

Example:

1) Simplify the expressions. $(4x^3 + 3x^4) - (x^4 - 5x^3) =$

First use Distributive Property for $-(x^4 - 5x^3)$, $\rightarrow -(x^4 - 5x^3) = -x^4 + 5x^3$

$(4x^3 + 3x^4) - (x^4 - 5x^3) = 4x^3 + 3x^4 - x^4 + 5x^3$

Now combine like terms: $4x^3 + 3x^4 - x^4 + 5x^3 = 2x^4 + 9 \quad ^3$

2) Add expressions. $(2x^3 - 6) + (9x^3 - 4x^2) =$

Remove parentheses: $(2x^3 - 6) + (9x^3 - 4x^2) = 2x^3 - 6 + 9x^3 - 4x^2$

Now combine like terms: $2x^3 - 6 + 9x^3 - 4x^2 = 11x^3 - 4x^2 - 6$

✍ *Add or subtract expressions.*

1) $(-x^2 - 2) + (2x^2 + 1) =$

2) $(2x^2 + 3) - (3 - 4x^2) =$

3) $(2x^3 + 3x^2) - (x^3 + 8) =$

4) $(4x^3 - x^2) + (3x^2 - 5x) =$

5) $(7x^3 + 9x) - (3x^3 + 2) =$

6) $(2x^3 - 2) + (2x^3 + 2) =$

7) $(4x^3 + 5) - (7 - 2x^3) =$

8) $(4x^2 + 2x^3) - (2x^3 + 5) =$

9) $(4x^2 - x) + (3x - 5x^2) =$

10) $(7x + 9) - (3x + 9) =$

11) $(4x^4 - 2x) - (6x - 2x^4) =$

12) $(12x - 4x^3) - (8x^3 + 6x) =$

Answers – Day 20

Writing Polynomials in Standard Form

1) $2x$
2) -3
3) $-5x^3 + 3x^{-2}$
4) $4x^3$
5) $-6x^3 + 2x^2 + x$
6) $2x^3 - x^2$

7) $4x^3 - 2x^2 + 2x$
8) $-6x^3 - 2x^2 + 4x$
9) $2x^2 - 5x + 2$
10) $9x^4 - 7x + 12$
11) $-2x^3 + 5x^2 + 13x$
12) $-x^3 + 6x^2 + 10$

Simplifying Polynomials

1) $10x - 50$
2) $8x^2 - 4x$
3) $20x^2 - 12x$
4) $21x^2 + 9x$
5) $32x^2 - 16x$
6) $25x^2 + 20x$

7) $2x^2 - 11x + 12$
8) $3x^2 - 11x - 20$
9) $x^2 - 8x + 15$
10) $9x^2 - 64$
11) $9x^2 - 36x + 32$
12) $-2x^3 + 6x^2$

Adding and Subtracting Polynomials

1) $x^2 - 1$
2) $6x^2$
3) $x^3 + 3x^2 - 8$
4) $4x^3 + 2x^2 - 5x$
5) $4x^3 + 9x - 2$
6) $4x^3$

7) $6x^3 - 2$
8) $4x^2 - 5$
9) $-x^2 + 2x$
10) $4x$
11) $6x^4 - 8x$
12) $-12x^3 + 6x$

Day 21:
Monomials Operations

Math Topics that you'll learn today:

✓ Multiplying Monomials

✓ Multiplying and Dividing Monomials

✓ Multiplying a Polynomial and a Monomial

Mathematics is, I believe, the chief source of the belief in eternal and exact truth, as well as a sensible intelligible world. – Bertrand Russell

Multiplying Monomials

Step-by-step guide:

✓ A monomial is a polynomial with just one term, like $2x$ or $7y$.

Example:

1) Multiply expressions. $5a^4b^3 \times 2a^3b^2 =$

Use this formula: $x^a \times x^b = x^{a+b}$

$a^4 \times a^3 = a^{4+3} = a^7$ and $b^3 \times b^2 = b^{3+2} = b^5$

Then: $5a^4b^3 \times 2a^3b^2 = 10a^7b^5$

2) Multiply expressions. $-4xy^4z^2 \times 3x^2y^5z^3 =$

Use this formula: $x^a \times x^b = x^{a+b}$

$x \times x^2 = x^{1+2} = x^3$, $y^4 \times y^5 = y^{4+5} = y^9$ and $z^2 \times z^3 = z^{2+3} = z^5$

Then: $-4xy^4z^2 \times 3x^2y^5z^3 = -12x^3y^9z^5$

✎ **_Simplify each expression._**

1) $4u^9 \times (-2u^3) =$

2) $(-2p^7) \times (-3p^2) =$

3) $3xy^2z^3 \times 2z^2 =$

4) $5u^5t \times 3ut^2 =$

5) $(-9a^6) \times (-5a^2b^4) =$

6) $-2a^3b^2 \times 4a^2b =$

7) $2xy^2 \times x^2y^3 =$

8) $3p^2q^4 \times (-2pq^3) =$

9) $4s^5t^2 \times 4st^3 =$

10) $(-6x^3y^2) \times 3x^2y =$

11) $2xy^2z \times 4z^2 =$

12) $4xy \times x^2y =$

Multiplying and Dividing Monomials

Step-by-step guide:

- ✓ When you divide two monomials you need to divide their coefficients and then divide their variables.
- ✓ In case of exponents with the same base, you need to subtract their powers.
- ✓ Exponent's rules:

$$x^a \times x^b = x^{a+b}, \qquad \frac{x^a}{x^b} = x^{a-b}$$
$$\frac{1}{x^b} = x^{-b}, \quad (x^a)^b = x^{a \times b}$$
$$(xy)^a = x^a \times y^a$$

Example:

1) Multiply expressions. $(-3x^7)(4x^3) =$
Use this formula: $x^a \times x^b = x^{a+b} \rightarrow x^7 \times x^3 = x^{10}$
Then: $(-3x^7)(4x^3) = -12x^{10}$

2) Dividing expressions. $\frac{18\ ^2y^5}{2xy^4} =$
Use this formula: $\frac{x^a}{x^b} = x^{a-b}$, $\frac{x^2}{x} = x^{2-1} = x$ and $\frac{y^5}{y^4} = y^{5-4} = y$
Then: $\frac{18x^2y^5}{2xy^4} = 9xy$

✎ *Simplify each expression.*

1) $(-2x^3y^4)(3x^3y^2) =$

2) $(-5x^3y^2)(-2x^4y^5) =$

3) $(9x^5y)(-3x^3y^3) =$

4) $(8x^7y^2)(6x^5y^4) =$

5) $(7x^4y^6)(4x^3y^4) =$

6) $(12x^2y^9)(7x^9y^{12}) =$

7) $\frac{12x^6y^8}{4x^4y^2} =$

8) $\frac{26x^9y^5}{2x^3y^4} =$

9) $\frac{80x^{12}y^9}{10\ ^6y^7} =$

10) $\frac{95\ ^{18}y^7}{5x^9y^2} =$

11) $\frac{200x^3y^8}{40x^3y^7} =$

12) $\frac{-15x^{17}y^{13}}{3x^6y^9} =$

Multiplying a Polynomial and a Monomial

Step-by-step guide:

✓ When multiplying monomials, use the product rule for exponents.

✓ When multiplying a monomial by a polynomial, use the distributive property.

$$a \times (b + c) = a \times b + a \times c$$

Example:

1) Multiply expressions. $-4x(5x + 9) =$

Use Distributive Property: $-4x(5x + 9) = -20x^2 - 36x$

2) Multiply expressions. $2x(6x^2 - 3y^2) =$

Use Distributive Property: $2x(6x^2 - 3y^2) = 12x^3 - 6xy^2$

✎ *Find each product.*

1) $3x(9x + 2y) =$

2) $6x(x + 2y) =$

3) $9x(2x + 4y) =$

4) $12x(3x + 9) =$

5) $11x(2x - 11y) =$

6) $2x(6x - 6y) =$

7) $2x(3x - 6y + 3) =$

8) $5x(3x^2 + 2y^2) =$

9) $13x(4x + 8y) =$

10) $5(2x^2 - 9y^2) =$

11) $3x(-2x^2y + 3y) =$

12) $-2(2x^2 - 2xy + 2) =$

Answers – Day 21

Multiplying Monomials

1) $-8u^{12}$
2) $6p^9$
3) $6xy^2z^5$
4) $15u^6t^3$
5) $45a^8b^4$
6) $-8a^5b^3$

7) $2x^3y^5$
8) $-6p^3q^7$
9) $16s^6t^5$
10) $-18x^5y^3$
11) $8xy^2z^3$
12) $4x^3y^2$

Multiplying and Dividing Monomials

1) $-6x^6y^6$
2) $10x^7y^7$
3) $-27x^8y^4$
4) $48x^{12}y^6$
5) $28x^7y^{10}$
6) $84x^{11}y^{21}$

7) $3x^2y^6$
8) $13x^6y$
9) $8x^6y^2$
10) $19x^9y^5$
11) $5y$
12) $-5x^{11}y^4$

Multiplying a Polynomial and a Monomial

1) $27x^2 + 6xy$
2) $6x^2 + 12xy$
3) $18x^2 + 36xy$
4) $36x^2 + 108x$
5) $22x^2 - 121xy$
6) $12x^2 - 12xy$

7) $6x^2 - 12xy + 6x$
8) $15x^3 + 10xy^2$
9) $52x^2 + 104xy$
10) $10x^2 - 45y^2$
11) $-6x^3y + 9xy$
12) $-4x^2 + 4xy - 4$

Day 22:
Polynomials Operations

Math Topics that you'll learn today:

- ✓ Multiplying Binomials

- ✓ Factoring Trinomials

- ✓ Operations with Polynomials

Mathematics compares the most diverse phenomena and discovers the secret analogies that unite them. ~ Joseph Fourier

Multiplying Binomials

Step-by-step guide:

- ✓ Use "FOIL". (First–Out–In–Last)

$$(x + a)(x + b) = x^2 + (b + a)x + ab$$

Example:

1) Multiply Binomials. $(x - 2)(x + 2) =$

Use "FOIL". (First–Out–In–Last): $(x - 2)(x + 2) = x^2 + 2x - 2x - 4$

Then simplify: $x^2 + 2x - 2x - 4 = ^2 - 4$

2) Multiply Binomials. $(x + 5)(x - 2) =$

Use "FOIL". (First–Out–In–Last):

$(x + 5)(x - 2) = x^2 - 2x + 5x - 10$

Then simplify: $x^2 - 2x + 5x - 10 = x^2 + 3x - 10$

✍ *Find each product.*

1) $(x + 2)(x + 2) =$

2) $(x - 3)(x + 2) =$

3) $(x - 2)(x - 4) =$

4) $(x + 3)(x + 2) =$

5) $(x - 4)(x - 5) =$

6) $(x + 5)(x + 2) =$

7) $(x - 6)(x + 3) =$

8) $(x - 8)(x - 4) =$

9) $(x + 2)(x + 8) =$

10) $(x - 2)(x + 4) =$

11) $(x + 4)(x + 4) =$

12) $(x + 5)(x + 5) =$

Factoring Trinomials

Step-by-step guide:

- ✓ "FOIL":
$$(x + a)(x + b) = x^2 + (b + a)x + ab$$
- ✓ "Difference of Squares":
$$a^2 - b^2 = (a + b)(a - b)$$
$$a^2 + 2ab + b^2 = (a + b)(a + b)$$
$$a^2 - 2ab + b^2 = (a - b)(a - b)$$
- ✓ "Reverse FOIL":
$$x^2 + (b + a)x + ab = (x + a)(x + b)$$

Example:

1) Factor this trinomial. $x^2 - 2x - 8 =$
   Break the expression into groups: $(x^2 + 2x) + (-4x - 8)$
   Now factor out x from $x^2 + 2x : x(x + 2)$ and factor out -4 from $-4x - 8: -4(x + 2)$
   Then: $= x(x + 2) - 4(x + 2)$, now factor out like term: $x + 2$
   Then: $(x + 2)(x - 4)$

2) Factor this trinomial. $x^2 - 6x + 8 =$
   Break the expression into groups: $(x^2 - 2x) + (-4x + 8)$
   Now factor out x from $x^2 - 2x : x(x - 2)$, and factor out -4 from $-4x + 8: -4(x - 2)$
   Then: $= x(x - 2) - 4(x - 2)$, now factor out like term: $x - 2$
   Then: $(x - 2)(x - 4)$

✍ *Factor each trinomial.*

1) $x^2 + 8x + 15 =$

2) $x^2 - 5x + 6 =$

3) $x^2 + 6x + 8 =$

4) $x^2 - 8x + 16 =$

5) $x^2 - 7x + 12 =$

6) $x^2 + 11x + 18 =$

7) $x^2 + 2x - 24 =$

8) $x^2 + 4x - 12 =$

9) $x^2 - 10x + 9 =$

10) $x^2 + 5x - 14 =$

11) $x^2 - 6x - 27 =$

12) $x^2 - 11x - 42 =$

Operations with Polynomials

Step-by-step guide:

    ✓ When multiplying a monomial by a polynomial, use the distributive property.

$$a \times (b + c) = a \times b + a \times$$

Example:

1) Multiply. $5(2x - 6) =$

Use the distributive property: $5(2x - 6) = 10x - 30$

2) Multiply. $2x(6x + 2) =$

Use the distributive property: $2x(6 + 2) = 12x^2 + 4x$

✎ *Find each product.*

1) $9(6x + 2) =$

2) $8(3x + 7) =$

3) $5(6x - 1) =$

4) $-3(8x - 3) =$

5) $3x^2(6x - 5) =$

6) $5x^2(7x - 2) =$

7) $6x^3(-3x + 4) =$

8) $-7x^4(2x - 4) =$

9) $8(x^2 + 2x - 3) =$

10) $4(4x^2 - 2x + 1) =$

11) $2(3x^2 + 2x - 2) =$

12) $8x(5x^2 + 3x + 8) =$

Answers – Day 22

Multiplying Binomials

1) $x^2 + 4x + 4$

2) $x^2 - x - 6$

3) $x^2 - 6x + 8$

4) $x^2 + 5x + 6$

5) $x^2 - 9x + 20$

6) $x^2 + 7x + 10$

7) $x^2 - 3x - 18$

8) $x^2 - 12x + 32$

9) $x^2 + 10x + 16$

10) $x^2 + 2x - 8$

11) $x^2 + 8x + 16$

12) $x^2 + 10x + 25$

Factoring Trinomials

1) $(x + 3)(x + 5)$

2) $(x - 2)(x - 3)$

3) $(x + 4)(x + 2)$

4) $(x - 4)(x - 4)$

5) $(x - 3)(x - 4)$

6) $(x + 2)(x + 9)$

7) $(x + 6)(x - 4)$

8) $(x - 2)(x + 6)$

9) $(x - 1)(x - 9)$

10) $(x - 2)(x + 7)$

11) $(x - 9)(x + 3)$

12) $(x + 3)(x - 14)$

Operations with Polynomials

1) $54x + 18$

2) $24x + 56$

3) $30x - 5$

4) $-24x + 9$

5) $18x^3 - 15x^2$

6) $35x^3 - 10x^2$

7) $-18x^4 + 24x^3$

8) $-14x^5 + 28x^4$

9) $8x^2 + 16x - 24$

10) $16x^2 - 8x + 4$

11) $6x^2 + 4x - 4$

12) $40x^3 + 24x^2 + 64x$

Day 23:

System of Equations

Math Topics that you'll learn today:

- ✓ Solving Systems of Equations

- ✓ Systems of Equations Word Problems

Mathematics is a hard thing to love. It has the unfortunate habit, like a rude dog, of turning its most unfavorable side towards you when you first make contact with it. ~ David Whiteland

Systems of Equations

Step-by-step guide:

- ✓ A system of equations contains two equations and two variables. For example, consider the system of equations: $x - y = 1, x + y = 5$
- ✓ The easiest way to solve a system of equation is using the elimination method. The elimination method uses the addition property of equality. You can add the same value to each side of an equation.
- ✓ For the first equation above, you can add $x + y$ to the left side and 5 to the right side of the first equation: $x - y + (x + y) = 1 + 5$. Now, if you simplify, you get: $x - y + (x + y) = 1 + 5 \rightarrow 2x = 6 \rightarrow x = 3$. Now, substitute 3 for the x in the first equation: $3 - y = 1$. By solving this equation, $y = 2$

Example:

What is the value of $x + y$ in this system of equations? $\begin{cases} 3x - 4y = -20 \\ -x + 2y = 10 \end{cases}$

Solving Systems of Equations by Elimination: $\begin{array}{l} 3x - 4y = -20 \\ -x + 2y = 10 \end{array}$ $\Rightarrow$ Multiply the second equation by 3, then add it to the first equation.

$\begin{array}{l} 3x - 4y = -20 \\ 3(-x + 2y = 10) \end{array} \Rightarrow \begin{array}{l} 3x - 4y = -20 \\ -3x + 6y = 30 \end{array}) \Rightarrow 2y = 10 \Rightarrow y = 5$. Now, substitute 5 for y in the first equation and solve for x. $3x - 4(5) = -20 \rightarrow 3x - 20 = -20 \rightarrow x = 0$

✎ *Solve each system of equations.*

1) $-2x + 2y = 4$ $x = $ ____

    $-2x + y = 3$ $y = $ ____

2) $-10x + 2y = -6$ $x = $ ____

    $6x - 16y = 48$ $y = $ ____

3) $y = -8$ $x = $ ____

    $16x - 12y = 32$

4) $2y = -6x + 10$ $x = $ ____

    $10x - 8y = -6$ $y = $ ____

5) $10x - 9y = -13$ $x = $ ____

    $-5x + 3y = 11$ $y = $ ____

6) $-3x - 4y = 5$ $x = $ ____

    $x - 2y = 5$ $y = $ ____

Systems of Equations Word Problems

Step-by-step guide:

✓ Define your variables, write two equations, and use elimination method for solving systems of equations.

Example:

Tickets to a movie cost $8 for adults and $5 for students. A group of friends purchased **20** tickets for $**115.00**. How many adults ticket did they buy? ____

Let x be the number of adult tickets and y be the number of student tickets. There are 20 tickets. Then: $x + y = 20$. The cost of adults' tickets is $8 and for students it is $5, and the total cost is $115. So, $8x + 5y = 20$. Now, we have a system of equations: $\begin{cases} x + y = 20 \\ 8x + 5y = 115 \end{cases}$

Multiply the first equation by -5 and add to the second equation: $-5(x + y = 20) = -5x - 5y = -100$

$8x + 5y + (-5x - 5y) = 115 - 100 \rightarrow 3x = 15 \rightarrow x = 5 \rightarrow 5 + y = 20 \rightarrow y = 15$. There are 5 adult tickets and 15 student tickets.

✍ *Solve each word problem.*

1) Tickets to a movie cost $5 for adults and $3 for students. A group of friends purchased **18** tickets for $**82.00**. How many adults ticket did they buy? _____

2) At a store, Eva bought two shirts and five hats for $**154.00**. Nicole bought three same shirts and four same hats for $**168.00**. What is the price of each shirt? _____

3) A farmhouse shelters **10** animals, some are pigs, and some are ducks. Altogether there are 36 legs. How many pigs are there? _____

4) A class of **195** students went on a field trip. They took **19** vehicles, some cars and some buses. If each car holds 5 students and each bus hold 25 students, how many buses did they take? _____

Answers – Day 23

Systems of Equations

1) $x = -1, y = 1$
2) $x = 0, y = -3$
3) $x = -4$
4) $x = 1, y = 2$
5) $x = -4, y = -3$
6) $x = 1, y = -2$

Systems of Equations Word Problems

1) 14
2) $32
3) 8
4) 5

Day 24:
Triangles and Polygons

Math Topics that you'll learn today:

- ✓ The Pythagorean Theorem

- ✓ Triangles

- ✓ Polygons

Mathematics is like checkers in being suitable for the young, not too difficult, amusing, and without peril to the state. ~ Plato

The Pythagorean Theorem

Step-by-step guide:

✓ In any right triangle: $a^2 + b^2 = c^2$

Example:

1) Find the missing length.

Use Pythagorean Theorem: $a^2 + b^2 = c^2$

Then: $a^2 + b^2 = c^2 \rightarrow 3^2 + 4^2 = c^2 \rightarrow 9 + 16 = c^2$

$c^2 = 25 \rightarrow c = 5$

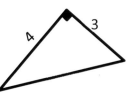

2) Right triangle ABC has two legs of lengths 6 cm (AB) and 8 cm (AC). What is the length of the third side (BC)?

Use Pythagorean Theorem: $a^2 + b^2 = c^2$

Then: $a^2 + b^2 = c^2 \rightarrow 6^2 + 8^2 = c^2 \rightarrow 36 + 64 = c^2$

$c^2 = 100 \rightarrow c = 10$

✎ *Find the missing side?*

1)

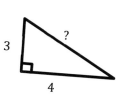

2)

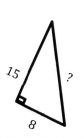

3)

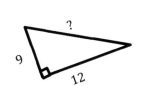

4)

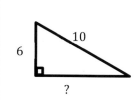

5)

6)

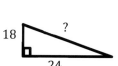

7)

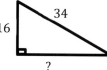

8)

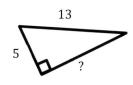

Triangles

Step-by-step guide:

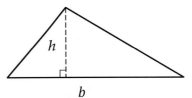

- ✓ In any triangle the sum of all angles is 180 degrees.
- ✓ Area of a triangle = $\frac{1}{2}$ (*base* × *height*)

Example:

What is the area of triangles?

1)

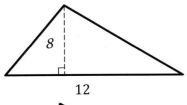

Solution:

Use the are formula: Area = $\frac{1}{2}$ (*base* × *height*)
base = 12 and *height* = 8
Area = $\frac{1}{2}(12 \times 8) = \frac{1}{2}(96) = 48$

2)

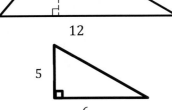

Solution:

Use the are formula: Area = $\frac{1}{2}$ (*base* × *height*)
base = 6 and *height* = 5
Area = $\frac{1}{2}(5 \times 6) = \frac{30}{2} = 15$

✍ *Find the measure of the unknown angle in each triangle.*

1)

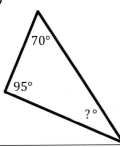

2)

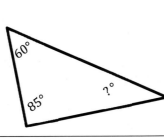

3)

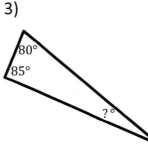

4)

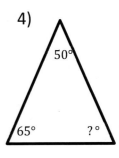

✍ *Find area of each triangle.*

5)

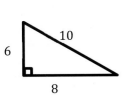

6)

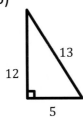

7)

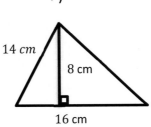

8)

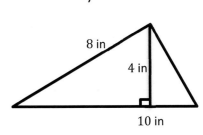

Polygons

Step-by-step guide:

Perimeter of a square $= 4 \times side = 4s$	Perimeter of a rectangle $= 2(width + length)$ 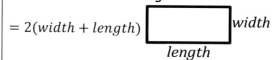
Perimeter of trapezoid $= a + b + c + d$	Perimeter of a regular hexagon $= 6a$
Example: Find the perimeter of following regular hexagon. 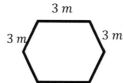 Perimeter of Pentagon $= 6a$ Perimeter of Pentagon $= 6a = 6 \times 3 = 18m$	Perimeter of a parallelogram $= 2(l + w)$ 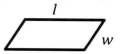

✎ *Find the perimeter of each shape.*

1)

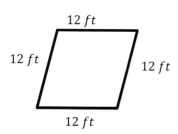

2)

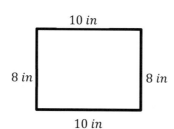

3)

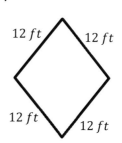

4)

14 cm

5) Regular hexagon

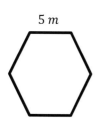

6)

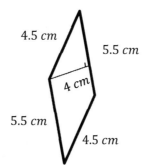

7) Parallelogram

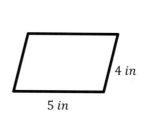

8) Square

6 m

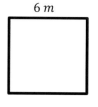

Answers – Day 24

The Pythagorean Theorem

1) 5

2) 17

3) 15

4) 8

5) 5

6) 30

7) 30

8) 12

Triangles

1) 15°

2) 35°

3) 15°

4) 65°

5) 24 *square unites*

6) 30 *square unites*

7) 64 *square unites*

8) 20 *square unites*

Polygons

1) 48 *ft*

2) 36 *in*

3) 48 *ft*

4) 56 *cm*

5) 30 *m*

6) 20 *cm*

7) 18 *in*

8) 24 *m*

Day 25:
Circles, Trapezoids and Cubes

Math Topics that you'll learn today:

- ✓ Circles

- ✓ Trapezoids

- ✓ Cubes

Mathematics is as much an aspect of culture as it is a collection of algorithms. ~ Carl Boyer

Circles

Step-by-step guide:

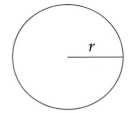

- ✓ In a circle, variable r is usually used for the radius and d for diameter and π is about 3.14.
- ✓ *Area of a circle* $= \pi r^2$
- ✓ *Circumference of a circle* $= 2\pi r$

Example:

1) Find the area of the circle.

Use area formula: $Area = \pi r^2$,

$r = 4$ then: $Area = \pi(4)^2 = 14\pi, \pi = 3.14$ then: $Area = 14 \times 3.14 = 43.96$

2) Find the Circumference of the circle.

Use Circumference formula: $Circumference = 2\pi r$

$r = 6$, then: $Circumference = 2\pi(6) = 12\pi$

$\pi = 3.14$ then: $Circumference = 12 \times 3.14 = 37.68$

✎ *Complete the table below.* ($\pi = 3.14$)

	Radius	Diameter	Circumference	Area
Circle 1	4 *inches*	8 *inches*	25.12 *inches*	50.24 *square inches*
Circle 2		12 *meters*		
Circle 3				12.56 *square ft*
Circle 4			18.84 miles	
Circle 5		5 *kilometers*		
Circle 6	6 *centimeters*			
Circle 7		8 *feet*		
Circle 8				28.26 *square meters*

Trapezoids

Step-by-step guide:

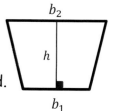

✓ A quadrilateral with at least one pair of parallel sides is a trapezoid.
✓ Area of a trapezoid $= \frac{1}{2}h(b_1 + b_2)$

Example:

Calculate the area of the trapezoid.

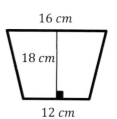

Use area formula: $A = \frac{1}{2}h(b_1 + b_2)$

$b_1 = 12$, $b_2 = 16$ and $h = 18$

Then: $A = \frac{1}{2}18(12 + 16) = 9(28) = 252\ cm^2$

✍ *Find the area of each trapezoid.*

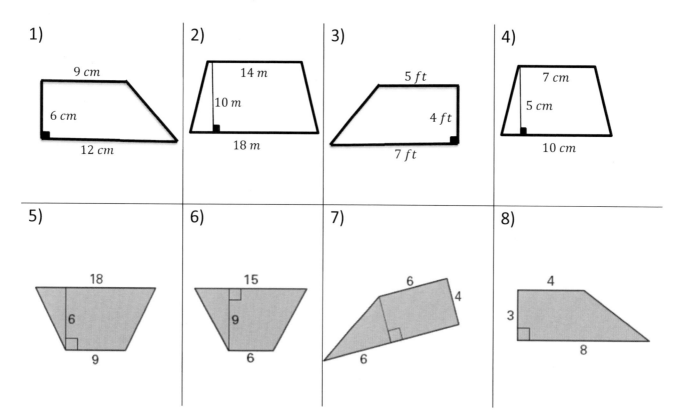

1)

9 cm

6 cm

12 cm

2)

14 m

10 m

18 m

3)

5 ft

4 ft

7 ft

4)

7 cm

5 cm

10 cm

5)

18

6

9

6)

15

9

6

7)

6

4

6

8)

4

3

8

Cubes

Step-by-step guide:

- ✓ A cube is a three-dimensional solid object bounded by six square sides.
- ✓ Volume is the measure of the amount of space inside of a solid figure, like a cube, ball, cylinder or pyramid.
- ✓ Volume of a cube $= (one\ side)^3$
- ✓ surface area of cube $= 6 \times (one\ side)^2$

Example:

Find the volume and surface area of this cube.

Use volume formula: $volume = (one\ side)^3$

Then: $volume = (one\ side)^3 = (2)^3 = 8\ cm^3$

Use surface area formula:

$surface\ area\ of\ cube: 6(one\ side)^2 = 6(2)^2 = 6(4) = 24\ cm^2$

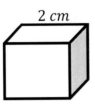
2 cm

✎ *Find the volume of each cube.*

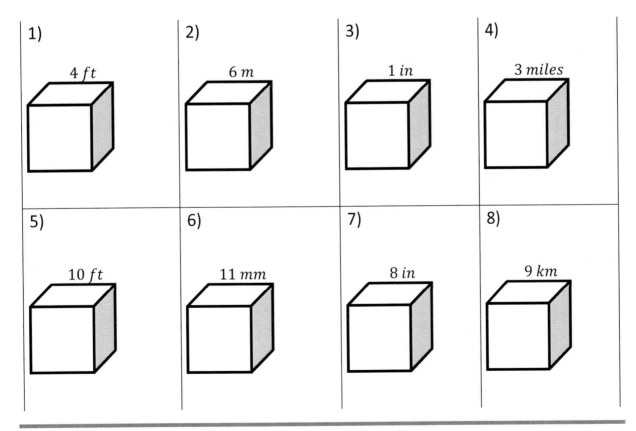

1) 4 ft

2) 6 m

3) 1 in

4) 3 miles

5) 10 ft

6) 11 mm

7) 8 in

8) 9 km

Answers – Day 25

Circles

	Radius	Diameter	Circumference	Area
Circle 1	4 *inches*	8 *inches*	25.12 *inches*	50.24 *square inches*
Circle 2	6 *meters*	12 *meters*	37.68 *meters*	113.04 *meters*
Circle 3	2 *square ft*	4 *square ft*	12.56 *square ft*	12.56 *square ft*
Circle 4	3 *miles*	6 *miles*	18.84 *miles*	28.26 *miles*
Circle 5	2.5 *kilometers*	5 *kilometers*	15.7 *kilometers*	19.63 *kilometers*
Circle 6	6 *centimeters*	12 *centimeters*	37.68 *centimeters*	113.04 *centimeters*
Circle 7	4 *feet*	8 *feet*	25.12 *feet*	50.24 *feet*
Circle 8	3 *square meters*	6 *square meters*	18.84 *square meters*	28.26 *square meters*

Trapezoids

1) $63\ cm^2$
2) $160\ m^2$
3) $24\ ft^2$
4) $42.5\ cm^2$

5) 81
6) 94.5
7) 36
8) 18

Cubes

1) $64\ ft^3$
2) $216\ m^3$
3) $1\ in^3$
4) $27\ miles^3$

5) $1,000\ ft^3$
6) $1,331\ mm^3$
7) $512\ in^3$
8) $729\ km^3$

138

Day 26:
Rectangular Prisms and Cylinder

Math Topics that you'll learn today:

✓ Rectangle Prisms

✓ Cylinder

It's fine to work on any problem, so long as it generates interesting mathematics along the way - even if you don't solve it at the end of the day." - Andrew Wiles

Rectangular Prisms

Step-by-step guide:

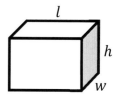

✓ A solid 3-dimensional object which has six rectangular faces.
✓ Volume of a Rectangular prism = **Length × Width × Height**

$Volume = l \times w \times h$ $Surface\ area = 2(wh + lw + lh)$

Example:

Find the volume and surface area of rectangular prism.

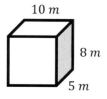

Use volume formula: $Volume = l \times w \times h$

Then: $Volume = 10 \times 5 \times 8 = 400\ m^3$

Use surface area formula: $Surface\ area = 2(wh + lw + lh)$

Then: $Surface\ area = 2(5 \times 8 + 10 \times 5 + 10 \times 8) = 2(40 + 50 + 80) = 340\ m^2$

 Find the volume of each Rectangular Prism.

1)

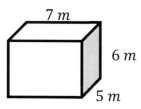

2)

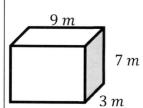

3)

9 m

7 m

3 m

4)

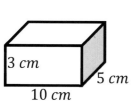

5)

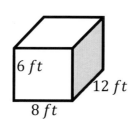

6)

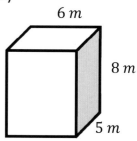

Cylinder

Step-by-step guide:

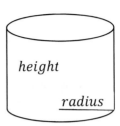

- ✓ A cylinder is a solid geometric figure with straight parallel sides and a circular or oval cross section.
- ✓ *Volume of Cylinder Formula* $= \pi(radius)^2 \times height$ $\pi = 3.14$
- ✓ *Surface area of a cylinder* $= 2\pi r^2 + 2\pi rh$

Example:

Find the volume and Surface area of the follow Cylinder.

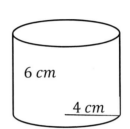

Use volume formula: $Volume = \pi(radius)^2 \times height$
Then: $Volume = \pi(4)^2 \times 6 = \pi 16 \times 6 = 96\pi$
$\pi = 3.14$ then: $Volume = 96\pi = 301.44$
Use surface area formula: $Surface\ area = 2\pi r^2 + 2\pi rh$
Then: $= 2\pi(4)^2 + 2\pi(4)(6) = 2\pi(16) + 2\pi(24) = 32\pi + 48\pi = 80\pi$
$\pi = 3.14$ then: $Surface\ area = 80 \times 3.14 = 251.2$

✍ *Find the volume of each Cylinder. Round your answer to the nearest tenth.* ($\pi = 3.14$)

1)

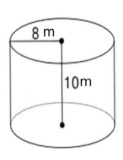

2)

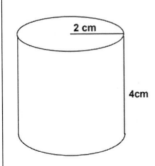

3)

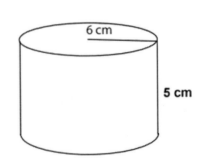

4)

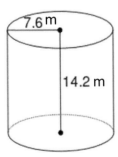

5)

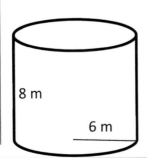

6)

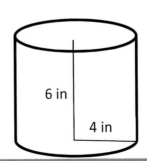

Answers – Day 26

Rectangle Prisms

1) $210 \ m^3$
2) $320 \ in^3$
3) $189 \ m^3$

4) $150 \ cm^3$
5) $576 \ ft^3$
6) $240 \ m^3$

Cylinder

1) $2,009.6 \ m^3$
2) $50.24 \ cm^3$
3) $565.2 \ cm^3$

4) $2,575.4 \ m^3$
5) $904.3 \ m^3$
6) $301.4 \ in^3$

Day 27:
Statistics

Math Topics that you'll learn today:

✓ Mean, Median, Mode, and Range of the Given Data

✓ Histograms

✓ Pie Graph

✓ Probability Problems

Millions saw the apple fall, but Newton asked why." - Bernard Baruch

Mean, Median, Mode, and Range of the Given Data

Step-by-step guide:

- ✓ Mean: $\dfrac{\text{sum of the data}}{\text{total number of data entires}}$
- ✓ Mode: value in the list that appears most often
- ✓ Range: the difference of largest value and smallest value in the list

Example:

1) What is the median of these numbers? $4, 9, 13, 8, 15, 18, 5$

   Write the numbers in order: $4, 5, 8, 9, 13, 15, 18$

   Median is the number in the middle. Therefore, the median is 9.

2) What is the mode of these numbers? $22, 16, 12, 9, 7, 6, 4, 6$

   Mode: value in the list that appears most often
   Therefore: mode is 6

✍ *Solve.*

1) In a javelin throw competition, five athletics score $56, 58, 63, 57$ and 61 meters. What are their Mean and Median? _____

2) Eva went to shop and bought 3 apples, 5 peaches, 8 bananas, 1 pineapple and 3 melons. What are the Mean and Median of her purchase? _____

✍ *Find Mode and Rage of the Given Data.*

3) $8, 2, 5, 9, 1, 2$

Mode: _____   Range: _____

4) $4, 4, 3, 9, 7, 9, 4, 6, 4$

Mode: _____   Range: _____

5) $6, 6, 2, 3, 6, 3, 9, 12$

Mode: _____   Range: _____

6) $12, 9, 2, 9, 3, 2, 9, 5$

Mode: _____   Range: _____

Histograms

Step-by-step guide:

✓ A histogram is an accurate representation of the distribution of numerical data.

Example:

Use the following Graph to complete the table.

Day	Distance (km)
1	
2	

Answer:

Day	Distance (km)
1	359
2	460
3	278
4	547
5	360

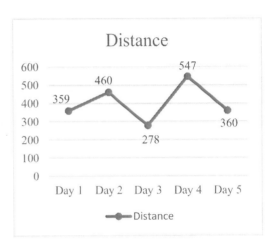

✍ The following table shows the number of births in the US from 2007 to 2012 (in millions).

Year	Number of births (in millions)
2007	4.32
2008	4.25
2009	4.13
2010	4
2011	3.95
2012	3.95

Draw a histogram for the table.

Pie Graph

Step-by-step guide:

✓ A Pie Chart is a circle chart divided into sectors, each sector represents the relative size of each value.

Example:

A library has 840 books that include Mathematics, Physics, Chemistry, English and History. Use following graph to answer question.

What is the number of Mathematics books?

Number of total books = 840,
Percent of Mathematics books = 30% = 0.30
Then: $0.30 \times 840 = 252$

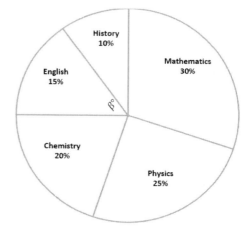

✍ **The circle graph below shows all Jason's expenses for last month. Jason spent $300 on his bills last month.**

1) How much did Jason spend on his car last month? _____

2) How much did Jason spend for foods last month? _____

3) How much did Jason spend on his rent last month? _____

4) What fraction is Jason's expenses for his bills and Car out of his total expenses last month?

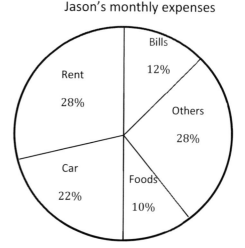

Jason's monthly expenses

Probability Problems

Step-by-step guide:

- ✓ Probability is the likelihood of something happening in the future. It is expressed as a number between zero (can never happen) to 1 (will always happen).
- ✓ Probability can be expressed as a fraction, a decimal, or a percent.

Example:

1) If there are 8 red balls and 12 blue balls in a basket, what is the probability that John will pick out a red ball from the basket?

There are 8 red ball and 20 are total number of balls. Therefore, probability that John will pick out a red ball from the basket is 8 out of 20 or $\frac{8}{8+1} = \frac{8}{20} = \frac{2}{5}$.

2) A bag contains 18 balls: two green, five black, eight blue, a brown, a red and one white. If 17 balls are removed from the bag at random, what is the probability that a brown ball has been removed?

If 17 balls are removed from the bag at random, there will be one ball in the bag.

The probability of choosing a brown ball is 1 out of 18. Therefore, the probability of not choosing a brown ball is 17 out of 18 and the probability of having not a brown ball after removing 17 balls is the same.

✍ *Solve.*

1) A number is chosen at random from 1 to 10. Find the probability of selecting number 4 or smaller numbers. _____

2) Bag A contains 9 red marbles and 3 green marbles. Bag B contains 9 black marbles and 6 orange marbles. What is the probability of selecting a green marble at random from bag A? What is the probability of selecting a black marble at random from Bag B? _____ _____

Answers – Day 27

Mean, Median, Mode, and Range of the Given Data

1) Mean: 59, Median: 58

2) Mean: 4, Median: 3

3) Mode: 2, Range: 8

4) Mode: 6, Range: 10

5) Mode: 4, Range: 6

6) Mode: 9, Range: 10

Histograms

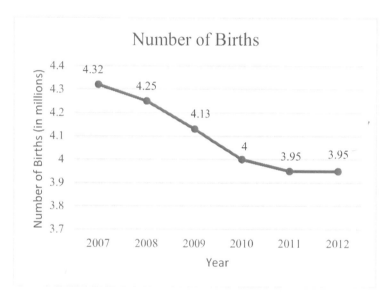

Pie Graph

1) $550
2) $250
3) $700
4) $\frac{17}{50}$

Probability Problems

1) $\frac{2}{5}$

2) $\frac{1}{4}, \frac{3}{5}$

Day 28:
Functions

Math Topics that you'll learn today:

- ✓ Function Notation
- ✓ Adding and Subtracting Functions
- ✓ Multiplying and Dividing Functions
- ✓ Composition of Functions

Mathematics is the supreme judge; from its decisions there is no appeal.

- Tobias Dantzig

Function Notation

Step-by-step guide:

- ✓ Functions are mathematical operations that assign unique outputs to given inputs.
- ✓ Function notation is the way a function is written. It is meant to be a precise way of giving information about the function without a rather lengthy written explanation.
- ✓ The most popular function notation is $f(x)$ which is read "f of x".

Example:

1) Evaluate: $w(x) = 3x + 1$, find $w(4)$. Substitute x with 4: Then: $w(x) = 3x + 1 \rightarrow w(4) = 3(4) + 1 \rightarrow w(x) = 12 + 1 \rightarrow w(x) = 13$

2) Evaluate: $h(n) = n^2 - 10$, find $h(-2)$. Substitute x with -2:

Then: $h(n) = n^2 - 10 \rightarrow h(-2) = (-2)^2 - 10 \rightarrow h(-2) = 4 - 10 \rightarrow h(-2) = -6$

✑ *Evaluate each function.*

1) $f(x) = -x + 5$, find $f(4)$

2) $g(n) = 10n - 3$, find $g(6)$

3) $g(n) = 8n + 4$, find $g(1)$

4) $h(x) = 4x - 22$, find $h(2)$

5) $h(a) = -11a + 5$, find $h(2a)$

6) $k(a) = 7a + 3$, find $k(a - 2)$

7) $h(x) = 3x + 5$, find $h(6x)$

8) $h(n) = n^2 - 10$, find $h(5)$

9) $h(n) = -2n^2 - 6n$, find $h(2)$

10) $g(n) = 3n^2 + 2n$, find $g(2)$

11) $h(x) = x^2 + 1$, find $h(\frac{x}{4})$

12) $h(x) = x^3 + 8$, find $h(-2)$

150

Adding and Subtracting Functions

Step-by-step guide:

✓ Just like we can add and subtract numbers, we can add and subtract functions. For example, if we had functions f and g, we could create two new functions:
✓ f + g and f - g.

Example:

1) $f(x) = 2x + 4, g(x) = x + 3$, **Find:** $(f - g)(1)$

$(f - g)(x) = f(x) - g(x)$, then: $(f - g)(x) = 2x + 4 - (x + 3) = 2x + 4 - x - 3 = x + 1$

Substitute x with 1: $(f - g)(1) = 1 + 1 = 2$

2) $g(a) = 2a - 1, f(a) = -a - 4$, **Find:** $(g + f)(-1)$

$(g + f)(a) = g(a) + f(a)$, Then: $(g + f)(a) = 2a - 1 - a - 4 = a - 5$

Substitute a with -1: $(g + f)(a) = a - 5 = -1 - 5 = -6$

✍ *Perform the indicated operation.*

1) $g(x) = 2x - 5$
   $h(x) = 4x + 5$
   Find: $g(3) - h(3)$

2) $h(3) = 3x + 3$
   $g(x) = -4x + 1$
   Find: $(h + g)(10)$

3) $f(x) = 4x - 3$
   $g(x) = x^3 + 2x$
   Find: $(f - g)(4)$

4) $h(n) = 4n + 5$
   $g(n) = 3n + 4$
   Find: $(h - g)(n)$

5) $g(x) = -x^2 - 1 - 2x$
   $f(x) = 5 + x$
   Find: $(g - f)(x)$

6) $g(t) = 2t + 5$
   $f(t) = -t^2 + 5$
   Find: $(g + f)(t)$

Multiplying and Dividing Functions

Step-by-step guide:

✓ Just like we can multiply and divide numbers, we can multiply and divide functions. For example, if we had functions f and g, we could create two new functions: f × g, and $\frac{f}{g}$.

Example:

1) $g(x) = -x - 2, f(x) = 2x + 1$, Find: $(g.f)(2)$

$(g.f)(x) = g(x).f(x) = (-x-2)(2x+1) = -2x^2 - x - 4x - 2 = -2x^2 - 5x - 2$

Substitute x with 2:

$(g.f)(x) = -2x^2 - 5x - 2 = -2(2)^2 - 5(2) - 2 = -8 - 10 - 2 = -20$

2) $f(x) = x + 4, h(x) = 5x - 2$, Find: $\left(\frac{f}{h}\right)(-1)$

$\left(\frac{f}{h}\right)(x) = \frac{f(x)}{h(x)} = \frac{x+4}{5x-2}$

Substitute x with -1: $\left(\frac{f}{h}\right)(x) = \frac{x+4}{5x-2} = \frac{(-1)+4}{5(-1)-2} = \frac{3}{-7} = -\frac{3}{7}$

✎ *Perform the indicated operation.*

1) $f(x) = 2a^2$
   $g(x) = -5 + 3a$
   Find $(\frac{f}{h})(2)$

2) $g(a) = 3a + 2$
   $f(a) = 2a - 4$
   Find $(\frac{g}{f})(3)$

3) $g(t) = t^2 + 3$
   $h(t) = 4t - 3$
   Find $(g.h)(-1)$

4) $g(n) = n^2 + 4 + 2n$
   $h(n) = -3n + 2$
   Find $(g.h)(1)$

5) $f(x) = 2x^3 - 5x^2$
   $g(x) = 2x - 1$
   Find $(f.g)(x)$

6) $f(x) = 3x - 1$
   $g(x) = x^2 - x$
   Find $(\frac{f}{g})(x)$

Composition of Functions

Step-by-step guide:

✓ The term "composition of functions" (or "composite function") refers to the combining together of two or more functions in a manner where the output from one function becomes the input for the next function.
✓ The notation used for composition is: $(f \circ g)(x) = f(g(x))$

Example:

1) **Using** $f(x) = x + 2$ **and** $g(x) = 4x$, **find:** $f(g(1))$

$(f \circ g)(x) = f(g(x))$

Then: $(f \circ g)(x) = f(g(x)) = f(4x) = 4x + 2$

Substitute x with 1: $(f \circ g)(1) = 4 + 2 = 6$

2) **Using** $f(x) = 5x + 4$ **and** $g(x) = x - 3$, **find:** $g(f(3))$

$(f \circ g)(x) = f(g(x))$

Then: $(g \circ f)(x) = g(f(x)) = g(5x + 4)$, *now substitute* x *in* g(x) *by* $5x + 4$. *Then:* $g(5x + 4) = (5x + 4) - 3 = 5x + 4 - 3 = 5x + 1$

Substitute x with 3: $(g \circ f)(x) = g(f(x)) = 5x + 1 = 5(3) + 1 = 15 = 1 = 16$

✍ **Using** $f(x) = 6x + 2$ **and** $g(x) = x - 5$, **find:**

1) $f(g(7))$ 3) $g(f(3))$

2) $f(f(2))$ 4) $g(g(x))$

✍ **Using** $f(x) = 7x + 4$ **and** $g(x) = 2x - 4$, **find:**

5) $f(g(3))$ 7) $g(f(4))$

6) $f(f(3))$ 8) $g(g(5))$

Answers – Day 28

Function Notation

1) 1
2) 57
3) 12
4) -8
5) $-22a + 5$
6) $7a - 11$
7) $18x + 5$

8) 0
9) -20
10) 16
11) $1 + \frac{1}{16}x^2$
12) 15

Adding and Subtracting Functions

1) -16
2) -6
3) -59

4) $n + 1$
5) $-x^2 - 3x - 6$
6) $-t^2 + 2t + 10$

Multiplying and Dividing Functions

1) $-\frac{2}{8} = -\frac{1}{4}$

2) $\frac{11}{2}$

3) -28

4) -7
5) $4x^4 - 12x^3 + 5x^2$
6) $\frac{3x-1}{x^2-x}$

Composition of functions

1) 14
2) 86
3) 15
4) $x - 10$

5) 18
6) 179
7) 60
8) 8

Day 29:

Time to Test

ISEE Upper Level Test Review

The Independent School Entrance Exam (ISEE) is a standardized test developed by the Educational Records Bureau for its member schools as part of their admission process.

There are currently four Levels of the ISEE:

- ✓ Primary Level (entering Grades 2 - 4)
- ✓ Lower Level (entering Grades 5 and 6)
- ✓ Middle Level (entering Grades 7 and 8)
- ✓ Upper Level (entering Grades 9 - 12)

There are five sections on the ISEE Upper Level Test:

- o Verbal Reasoning
- o Quantitative Reasoning
- o Reading Comprehension
- o Mathematics Achievement
- o and a 30-minute essay

ISEE Upper Level tests use a multiple-choice format and contain two Mathematics sections:

Quantitative Reasoning

There are 37 questions in the Quantitative Reasoning section and students have 35 minutes to answer the questions. This section contains word problems and quantitative comparisons. The word problems require either no calculation or simple calculation. The quantitative comparison items present two quantities, (A) and (B), and the student needs to select one of the following four answer choices:

(A) The quantity in Column A is greater.

(B) The quantity in Column B is greater.

(C) The two quantities are equal.

(D) The relationship cannot be determined from the information given.

Mathematics Achievement

There are 47 questions in the Mathematics Achievement section and students have 40 minutes to answer the questions. Mathematics Achievement measures students' knowledge of Mathematics requiring one or more steps in calculating the answer.

In this book, we have reviewed Quantitative Reasoning and Mathematic Achievement topics being tested on the ISEE Upper Level. In this section, there are two complete ISEE Upper Level Quantitative Reasoning and Mathematics Achievement Tests. Let your student take these tests to see what score they will be able to receive on a real ISEE Upper Level test.

Good luck!

Time to refine your skill with a practice examination

Take practice ISEE Upper Level Math Tests to simulate the test day experience. After you've finished, score your tests using the answer keys.

Before You Start

- You'll need a pencil and a timer to take the test.

- After you've finished the test, review the answer key to see where you went wrong.

- Use the answer sheet provided to record your answers. (You can cut it out or photocopy it)

- You will receive 1 point for every correct answer, and you will lose $\frac{1}{4}$ point for each incorrect answer. There is no penalty for skipping a question.

Calculators are NOT permitted for the ISEE Upper Level Test

Good Luck!

Mathematics is like love; a simple idea, but it can get complicated.

ISEE Upper Level Math

Practice Test 2019

Two Parts

Total number of questions: 84

Part 1 (Quantitative Reasoning): 37 questions

Part 2 (Mathematics Achievement): 47 questions

Total time for two parts: 75 Minutes

ISEE Upper Level Practice Test Answer Sheets

Remove (or photocopy) these answer sheets and use them to complete the practice tests.

ISEE Upper Level Practice Test
Quantitative Reasoning

1) Ⓐ Ⓑ Ⓒ Ⓓ   2) Ⓐ Ⓑ Ⓒ Ⓓ

3) Ⓐ Ⓑ Ⓒ Ⓓ   4) Ⓐ Ⓑ Ⓒ Ⓓ

5) Ⓐ Ⓑ Ⓒ Ⓓ   6) Ⓐ Ⓑ Ⓒ Ⓓ

7) Ⓐ Ⓑ Ⓒ Ⓓ   8) Ⓐ Ⓑ Ⓒ Ⓓ

9) Ⓐ Ⓑ Ⓒ Ⓓ   10) Ⓐ Ⓑ Ⓒ Ⓓ

11) Ⓐ Ⓑ Ⓒ Ⓓ   12) Ⓐ Ⓑ Ⓒ Ⓓ

13) Ⓐ Ⓑ Ⓒ Ⓓ   14) Ⓐ Ⓑ Ⓒ Ⓓ

15) Ⓐ Ⓑ Ⓒ Ⓓ   16) Ⓐ Ⓑ Ⓒ Ⓓ

17) Ⓐ Ⓑ Ⓒ Ⓓ   18) Ⓐ Ⓑ Ⓒ Ⓓ

19) Ⓐ Ⓑ Ⓒ Ⓓ   20) Ⓐ Ⓑ Ⓒ Ⓓ

21) Ⓐ Ⓑ Ⓒ Ⓓ   22) Ⓐ Ⓑ Ⓒ Ⓓ

23) Ⓐ Ⓑ Ⓒ Ⓓ   24) Ⓐ Ⓑ Ⓒ Ⓓ

25) Ⓐ Ⓑ Ⓒ Ⓓ   26) Ⓐ Ⓑ Ⓒ Ⓓ

27) Ⓐ Ⓑ Ⓒ Ⓓ   28) Ⓐ Ⓑ Ⓒ Ⓓ

29) Ⓐ Ⓑ Ⓒ Ⓓ   30) Ⓐ Ⓑ Ⓒ Ⓓ

31) Ⓐ Ⓑ Ⓒ Ⓓ   32) Ⓐ Ⓑ Ⓒ Ⓓ

33) Ⓐ Ⓑ Ⓒ Ⓓ   34) Ⓐ Ⓑ Ⓒ Ⓓ

35) Ⓐ Ⓑ Ⓒ Ⓓ   36) Ⓐ Ⓑ Ⓒ Ⓓ

37) Ⓐ Ⓑ Ⓒ Ⓓ

ISEE Upper Level Practice Test
Mathematics Achievement

1) Ⓐ Ⓑ Ⓒ Ⓓ	2) Ⓐ Ⓑ Ⓒ Ⓓ
3) Ⓐ Ⓑ Ⓒ Ⓓ	4) Ⓐ Ⓑ Ⓒ Ⓓ
5) Ⓐ Ⓑ Ⓒ Ⓓ	6) Ⓐ Ⓑ Ⓒ Ⓓ
7) Ⓐ Ⓑ Ⓒ Ⓓ	8) Ⓐ Ⓑ Ⓒ Ⓓ
9) Ⓐ Ⓑ Ⓒ Ⓓ	10) Ⓐ Ⓑ Ⓒ Ⓓ
11) Ⓐ Ⓑ Ⓒ Ⓓ	12) Ⓐ Ⓑ Ⓒ Ⓓ
13) Ⓐ Ⓑ Ⓒ Ⓓ	14) Ⓐ Ⓑ Ⓒ Ⓓ
15) Ⓐ Ⓑ Ⓒ Ⓓ	16) Ⓐ Ⓑ Ⓒ Ⓓ
17) Ⓐ Ⓑ Ⓒ Ⓓ	18) Ⓐ Ⓑ Ⓒ Ⓓ
19) Ⓐ Ⓑ Ⓒ Ⓓ	20) Ⓐ Ⓑ Ⓒ Ⓓ
21) Ⓐ Ⓑ Ⓒ Ⓓ	22) Ⓐ Ⓑ Ⓒ Ⓓ
23) Ⓐ Ⓑ Ⓒ Ⓓ	24) Ⓐ Ⓑ Ⓒ Ⓓ
25) Ⓐ Ⓑ Ⓒ Ⓓ	26) Ⓐ Ⓑ Ⓒ Ⓓ
27) Ⓐ Ⓑ Ⓒ Ⓓ	28) Ⓐ Ⓑ Ⓒ Ⓓ
29) Ⓐ Ⓑ Ⓒ Ⓓ	30) Ⓐ Ⓑ Ⓒ Ⓓ
31) Ⓐ Ⓑ Ⓒ Ⓓ	32) Ⓐ Ⓑ Ⓒ Ⓓ
33) Ⓐ Ⓑ Ⓒ Ⓓ	34) Ⓐ Ⓑ Ⓒ Ⓓ
35) Ⓐ Ⓑ Ⓒ Ⓓ	36) Ⓐ Ⓑ Ⓒ Ⓓ
37) Ⓐ Ⓑ Ⓒ Ⓓ	38) Ⓐ Ⓑ Ⓒ Ⓓ
39) Ⓐ Ⓑ Ⓒ Ⓓ	40) Ⓐ Ⓑ Ⓒ Ⓓ
41) Ⓐ Ⓑ Ⓒ Ⓓ	42) Ⓐ Ⓑ Ⓒ Ⓓ
43) Ⓐ Ⓑ Ⓒ Ⓓ	44) Ⓐ Ⓑ Ⓒ Ⓓ
45) Ⓐ Ⓑ Ⓒ Ⓓ	46) Ⓐ Ⓑ Ⓒ Ⓓ
47) Ⓐ Ⓑ Ⓒ Ⓓ	

ISEE Upper Level Math Practice Test 2019

Section 1

Quantitative Reasoning

37 questions

Total time for this section: 35 Minutes

You may NOT use a calculator for this test.

1) If $2y + 6 < 30$, then y could be equal to?

A. 15

B. 14

C. 12

D. 8

2) Which of the following is NOT a factor of 90?

A. 9

B. 10

C. 16

D. 30

3) What is the area of a square whose diagonal is 6?

A. 20

B. 18

C. 12

D. 10

4) If Emily left a $13.26 tip on a breakfast that cost $58.56, approximately what percentage was the tip?

A. 24%

B. 22%

C. 20%

D. 18%

5) There are 7 blue marbles, 9 red marbles, and 6 yellow marbles in a box. If Ava randomly selects a marble from the box, what is the probability of selecting a red or yellow marble?

A. $\frac{1}{7}$

B. $\frac{1}{9}$

C. $\frac{15}{22}$

D. $\frac{5}{7}$

6) James earns $8.50 per hour and worked 20 hours. Jacob earns $10.00 per hour. How many hours would Jacob need to work to equal James's earnings over 20 hours?

A. 14

B. 17

C. 20

D. 25

7) A phone company charges $5 for the first five minutes of a phone call and 50 cents per minute thereafter. If Sofia makes a phone call that lasts 30 minutes, what will be the total cost of the phone call?

A. 18.00

B. 18.50

C. 20.00

D. 20.50

8) Michelle and Alec can finish a job together in 50 minutes. If Michelle can do the job by herself in 2.5 hours, how many minutes does it take Alec to finish the job?

A. 100

B. 75

C. 50

D. 40

9) If 150% of a number is 75, then what is the 90% of that number?

A. 45

B. 50

C. 60

D. 70

10) In the figure, MN is 50 cm. How long is ON?

A. 35 cm

B. 30 cm

C. 25 cm

D. 20 cm

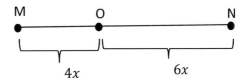

11) On a map, the length of the road from City A to City B is measured to be 18 inches. On this map, $\frac{1}{2}$ inch represents an actual distance of 14 miles. What is the actual distance, in miles, from City A to City B along this road?

A. 504
B. 620
C. 860
D. 1,260

A library has 840 books that include Mathematics, Physics, Chemistry, English and History. Use following graph to answer questions 12.

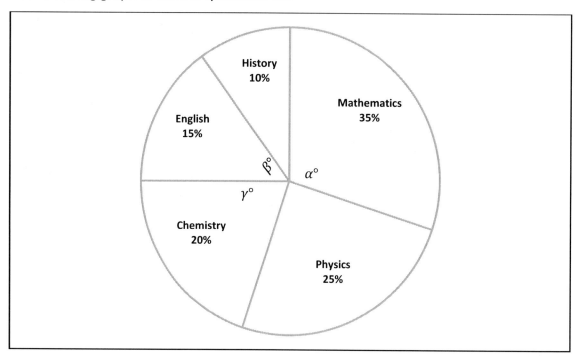

12) What is the product of the number of Mathematics and number of English books?

A. 42,168
B. 37,044
C. 29,460
D. 18,640

13) How many $\frac{1}{6}$ pound paperback books together weigh 60 pounds?

A. 100

B. 200

C. 300

D. 360

14) Emily and Daniel have taken the same number of photos on their school trip. Emily has taken 6 times as many as photos as Claire and Daniel has taken 15 more photos than Claire. How many photos has Claire taken?

A. 2

B. 3

C. 9

D. 11

15) The first four terms in a sequence are shown below. What is the sixth term in the sequence?

$\{3, 6, 11, 18, \dots\}$

A. 38

B. 40

C. 45

D. 50

16) What is the equation of the line that passes through $(3, -3)$ and has a slope of 7?

A. $y = 2x - 16$

B. $y = 3x - 24$

C. $y = 2x + 16$

D. $y = 3x + 24$

17) Find the solution (x, y) to the following system of equations?
$$-4x - y = 8$$
$$8x + 6y = 20$$

A. $(5, 14)$

B. $(8, 6)$

C. $(17, 11)$

D. $\left(-\frac{17}{4}, 9\right)$

18) Sophia purchased a sofa for $530.20. The sofa is regularly priced at $631. What was the percent discount Sophia received on the sofa?

A. 12%

B. 16%

C. 21%

D. 26%

19) Three second of 20 is equal to $\frac{5}{2}$ of what number?

A. 12

B. 20

C. 40

D. 60

20) A supermarket's sales increased by 11 percent in the month of April and decreased by 11 percent in the month of May. What is the percent change in the sales of the supermarket over the two-month period?

A. 2% decrease

B. No change

C. 2% increase

D. 2.2% increase

21) The distance between cities A and B is approximately 2,700 miles. If Alice drives an average of 74 miles per hour, how many hours will it take Alice to drive from city A to city B?

A. *Approximately* 41 *hours*

B. *Approximately* 36 *hours*

C. *Approximately* 28 *hours*

D. *Approximately* 21 *hours*

Quantitative Comparisons

Direction: Questions 22 to 37 are Quantitative Comparisons Questions. Using the information provided in each question, compare the quantity in column A to the quantity in Column B. Choose on your answer sheet grid

- A   if the quantity in Column A is greater
- B   if the quantity in Column B is greater
- C   if the two quantities are equal
- D   if the relationship cannot be determined from the information given

22)

Column A	Column B
The sum of all members in Set A	2

23) Set A includes even prime numbers.

Column A	Column B
5%	$\dfrac{1}{2}$

24)

Column A	Column B
The average of $12, 18, 26, 30, 32$	The average of $24, 28, 30, 36$

‍

25)

Column A	Column B
The number of posts needed for a fence 100 feet long if the posts are placed 12.5 feet apart	8 posts

26) $\frac{x}{4} = y^2$

Column A	Column B
x	y

27) $x = -1$

Column A	Column B
$3x^2 - 2x + 4$	$2x^3 + x^2 + 4$

28)

Column A	Column B
$9 + 12(9 - 5)$	$12 + 9(9 - 5)$

29) $\frac{x}{48} = \frac{2}{3}$

Column A	Column B
$\dfrac{8}{x}$	$\dfrac{1}{4}$

30) Working at constant rates, machine D makes b rolls of steel in 25 minutes and machine E makes b rolls of steel in one hour ($b > 0$)

Column A	Column B
The number of rolls of steel made by machine D in 2 hours and 5 minutes.	The number of rolls of steel made by machine E in 4 hours.

31) $6 > y > -2$

Column A	Column B
$\dfrac{y}{4}$	$\dfrac{4}{y}$

32) $\dfrac{a}{b} = \dfrac{c}{d}$

Column A	Column B
$a + b$	$c + d$

33) The ratio of boys to girls in a class is 7 to 11.

Column A	Column B
Ratio of boys to the entire class	$\dfrac{1}{3}$

34) There are 6 blue marbles and 4 green marbles in a jar. Two marbles are pulled out in succession without replacing them in the jar.

Column A	Column B
The probability that both marbles are blue.	The probability that the first marbles is green, but the second is blue.

35) A magazine printer consecutively numbered the pages of a magazine, starting with 1 on the first page, 10 on the tenth page, etc. In numbering the gages, the printer printed a total of 195 digits.

Column A	Column B
The number of pages in the magazine	100

36)

Column A	Column B
The largest number that can be written	The largest number that can be written
by rearranging the digits in 263	by rearranging the digits in 192

37) A computer priced $124 includes 4% profit

Column A	Column B
$119	The original cost of the computer

IF YOU FINISH BEFORE TIME IS CALLED, YOU MAY CHECK YOUR WORK ON THIS SECTION ONLY. DO NOT TURN TO ANY OTHER SECTION IN THE TEST. **STOP**

ISEE Upper Level Math Practice Test 2019

Section 2

Mathematics Achievement

47 questions

Total time for this section: 40 Minutes

You may NOT use a calculator for this test.

1) How is this number written in scientific notation?

$$0.00003379$$

A. 3.379×10^{-5}

B. 33.79×10^{6}

C. 0.3379×10^{-4}

D. 3379×10^{-8}

2) $|10 - (12 \div |1 - 5|)| = ?$

A. 7

B. -7

C. 5

D. -5

3) $(x + 4)(x + 5) =$

A. $x^2 + 9x + 20$

B. $2x + 12x + 12$

C. $x^2 + 25x + 10$

D. $x^2 + 12x + 35$

4) Find all values of x for which $6x^2 + 16x + 8 = 0$

A. $-\dfrac{3}{2}, -\dfrac{1}{2}$

B. $-\dfrac{2}{3}, -2$

C. $-2, -\dfrac{1}{4}$

D. $-\dfrac{2}{3}, \dfrac{1}{2}$

5) Which of the following graphs represents the compound inequality $-2 \le 2x - 4 < 8$?

A.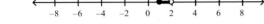

B.

C.

D.

6) $1 - 9 \div (4^2 \div 2) = $ ___

A. 6

B. $\dfrac{3}{4}$

C. -1

D. -2

7) A girl $200\ cm$ tall, stands $460\ cm$ from a lamp post at night. Her shadow from the light is $80\ cm$ long. How high is the lamp post?

A. 440

B. 500

C. 900

D. 1350

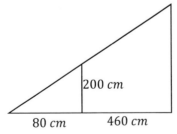

200 cm

80 cm   460 cm

8) Which graph shows a non-proportional linear relationship between x and y?

A.

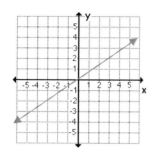

B.

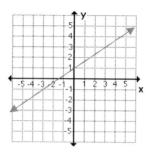

C.

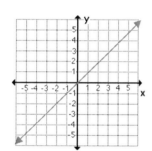

D.

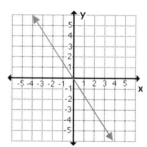

9) The ratio of boys to girls in a school is $3:5$. If there are 600 students in a school, how many boys are in the school?

A. 200

B. 225

C. 300

D. 340

10) $90 \div \frac{1}{9} = ?$

A. 9.125

B. 10

C. 81

D. 810

11) The rectangle on the coordinate grid is translated 5 units down and 4 units to the left.

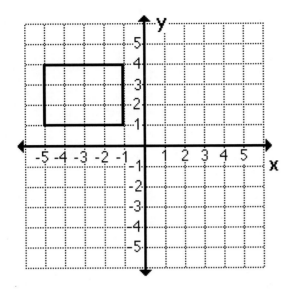

Which of the following describes this transformation?

A. $(x, y) \Rightarrow (x - 4, y + 5)$
B. $(x, y) \Rightarrow (x - 4, y - 5)$
C. $(x, y) \Rightarrow (x + 4, y + 5)$
D. $(x, y) \Rightarrow (x + 4, y - 5)$

12) Find the area of a rectangle with a length of 148 feet and a width of 90 feet.

A. $13,320 \ sq.\ ft$

B. $13,454 \ sq.\ ft$

C. $13,404 \ sq.\ ft$

D. $13,204 \ sq.\ ft$

13) Which value of x makes the following inequality true?

$$\frac{4}{23} \leq x < 25\%$$

A. 0.12

B. $\frac{5}{36}$

C. $\sqrt{0.044}$

D. 0.104

14) When an integer is multiplied by itself, it can end in all of the following digits EXCEPT

A. 1

B. 6

C. 8

D. 9

15) If the area of trapezoid is $126\ cm$, what is the perimeter of the trapezoid?

A. $12\ cm$

B. $32\ cm$

C. $46\ cm$

D. $55\ cm$

16) Emily lives $4\frac{1}{5}$ miles from where she works. When traveling to work, she walks to a bus stop $\frac{1}{2}$ of the way to catch a bus. How many miles away from her house is the bus stop?

A. $2\frac{1}{10}\ miles$

B. $4\frac{3}{10}\ miles$

C. $2\frac{3}{10}\ miles$

D. $1\frac{3}{10}\ miles$

17) If a vehicle is driven 40 miles on Monday, 45 miles on Tuesday, and 50 miles on Wednesday, what is the average number of miles driven each day?

A. $40 \; miles$

B. $45 \; miles$

C. $50 \; miles$

D. $53 \; miles$

18) Use the diagram below to answer the question.

Given the lengths of the base and diagonal of the rectangle below, what is the length of height h, in terms of s?

A. $s\sqrt{2}$

B. $2s\sqrt{2}$

C. $3s$

D. $3s^2$

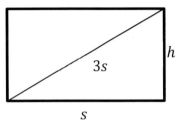

Use the chart below to answer the question.

Color	Number
White	40
Black	30
Beige	40

19) There are also purple marbles in the bag. Which of the following can NOT be the probability of randomly selecting a purple marble from the bag?

A. $\dfrac{1}{11}$

B. $\dfrac{1}{6}$

C. $\dfrac{2}{5}$

D. $\dfrac{1}{23}$

20) With an 23% discount, Ella was able to save $21.87 on a dress. What was the original price of the dress?

A. $89.92

B. $91.82

C. $95.08

D. $97.92

21) $\dfrac{8}{35}$ is equals to:

A. 0.3

B. 2.5

C. 0.04

D. 0.22

22) If 30% of A is 1,200, what is 12% of A?

A. 280

B. 480

C. 1,200

D. 1600

23) Simplify $\dfrac{\dfrac{1}{2} - \dfrac{x+5}{4}}{\dfrac{x^3}{2} - \dfrac{5}{2}}$

A. $\dfrac{3+x}{x^3+10}$

B. $\dfrac{3-x}{2x^3-10}$

C. $\dfrac{3+x}{x^3-10}$

D. $\dfrac{-3-x}{2x^3-10}$

24) If $(6.2 + 8.3 + 2.4) \times x = x$, then what is the value of x?

A. 0

B. $\dfrac{3}{10}$

C. -6

D. -12

25) Two dice are thrown simultaneously, what is the probability of getting a sum of 5 or 8?

A. $\dfrac{1}{3}$

B. $\dfrac{1}{4}$

C. $\dfrac{1}{16}$

D. $\dfrac{11}{36}$

26) If 7 garbage trucks can collect the trash of 38 homes in a day. How many trucks are needed to collect in 190 houses?

A. 20

B. 30

C. 35

D. 40

27) $78.56 \div 0.05 = ?$

A. 15.712

B. 1571.2

C. 157.12

D. 1.5712

28) In the following figure, AB is the diameter of the circle. What is the circumference of the circle?

A. 4π

B. 6π

C. 8π

D. 10π

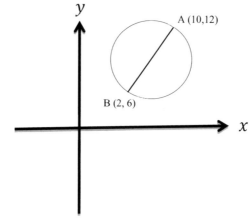

29) What is the value of x in the following equation?
$$2x^2 + 6 = 26$$

A. ± 4

B. $\pm \sqrt{9}$

C. $\pm \sqrt{10}$

D. ± 3

30) A circle has a diameter of 20 inches. What is its approximate area?

A. 314

B. 114

C. 74.00

D. 12.56

31) $6\ days\ 20\ hours\ 36\ minutes\ -\ 4\ days\ 12\ hours\ 24\ minutes\ =\ ?$

A. $2\ days\ 8\ hours\ and\ 12 minutes$

B. $1\ days\ 8\ hours\ 12\ minutes$

C. $2\ days\ 7\ hours\ 14\ minutes$

D. $1\ days\ 7\ hours\ 14\ minutes$

32) The base of a right triangle is 4 feet, and the interior angles are $45 - 45 - 90$. What is its area?

A. 2 *square feet*

B. 4 *square feet*

C. 8 *square feet*

D. 10 *square feet*

Use the following table to answer question below.

DANIEL'S BIRD-WATCHING PROJECT	
DAY	NUMBER OF RAPTORS SEEN
Monday	?
Tuesday	10
Wednesday	15
Thursday	13
Friday	6
MEAN	11

33) The above table shows the data Daniel collects while watching birds for one week. How many raptors did Daniel see on Monday?

A. 10

B. 11

C. 12

D. 13

34) A floppy disk shows 837,036 bytes free and 639,352 bytes used. If you delete a file of size 542,159 bytes and create a new file of size 499,986 bytes, how many free bytes will the floppy disk have?

A. 567,179

B. 671,525

C. 879,209

D. 899,209

35) Increased by 40%, the number 70 becomes:

A. 40

B. 98

C. 126

D. 130

36) If $12 + x^{\frac{1}{2}} = 24$, then what is the value of $5 \times x$?

A. 15

B. 60

C. 240

D. 720

37) Which equation represents the statement "twice the difference between 5 times H and 2 gives 35".

A. $\dfrac{5H + 2}{2} = 35$

B. $5(2H + 2) = 35$

C. $2(5H - 2) = 35$

D. $2\,\dfrac{5H}{2} = 35$

38) A circle is inscribed in a square, as shown below.

The area of the circle is $25\pi \ cm^2$ What is the area of the square?

A. $10cm^2$
B. $26cm^2$
C. $48 \ cm^2$
D. $100 \ cm^2$

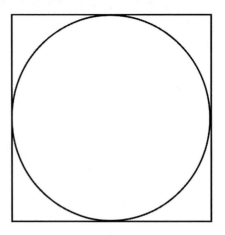

39) Triangle ABC is graphed on a coordinate grid with vertices at $A \ (-2, -3)$, $B \ (-4, 1)$ and $C \ (9, 7)$. Triangle ABC is reflected over x axes to create triangle $A'B'C'$.

Which order pair represents the coordinate of C'?

A. $(9, 7)$
B. $(-9, -7)$
C. $(9, -7)$
D. $(7, -9)$

40) Which set of ordered pairs represents y as a function of x?

A. $\{(5, -2), (5, 7), (9, -8), (4, -7)\}$
B. $\{(2, 2), (3, -9), (5, 8), (2, 7)\}$
C. $\{(9, 12), (8, 7), (6, 11), (8, 18)\}$
D. $\{(6, 1), (3, 1), (0, 5), (6, 1)\}$

41) The width of a box is one third of its length. The height of the box is one half of its width. If the length of the box is $24 \ cm$, what is the volume of the box?

A. $80cm^3$
B. $165 \ cm^3$
C. $243 \ cm^3$
D. $768cm^3$

42) How many 4×4 squares can fit inside a rectangle with a height of 40 and width of 12?

A. 60

B. 50

C. 40

D. 30

43) David makes a weekly salary of $230 plus 9% commission on his sales. What will his income be for a week in which he makes sales totaling $1,200?

A. $338

B. $318

C. $308

D. $298

44) $5x^3y^2 + 4x^5y^3 - (6x^3y^2 - 4x^1y^5) =$ ___

A. $4x^5y^3$

B. $4x^5y^3 - x^3y^2 + 4xy^5$

C. x^3y^2

D. $4x^5y^3 - x^3y^2$

45) The radius of circle A is five times the radius of circle B. If the circumference of circle A is 20π, what is the area of circle B?

A. 3π

B. 4π

C. 6π

D. 12π

46) A square measures 8 inches on one side. By how much will the area be decreased if its length is increased by 5 inches and its width decreased by 4 inches.

A. 1 sq decreased

B. 3 sq decreased

C. 8 sq decreased

D. 12 sq decreased

47) If a box contains red and blue balls in ratio of 3 : 2 red to blue, how many red balls are there if 80 blue balls are in the box?

A. 60

B. 80

C. 100

D. 120

IF YOU FINISH BEFORE TIME IS CALLED, YOU MAY CHECK YOUR WORK ON THIS SECTION.

STOP

ISEE Upper Level Math Practice Test Answers and Explanations

❋ Now, it's time to review your results to see where you went wrong and what areas you need to improve!

ISEE Upper Level Math Practice Test 1 Answer Key									
Quantitative Reasoning					**Mathematics Achievement**				
1	B	17	A	33	D	1	B	17	B	33	A
2	C	18	D	34	D	2	A	18	A	34	C
3	C	19	B	35	A	3	D	19	D	35	B
4	B	20	A	36	B	4	A	20	C	36	C
5	C	21	B	37	B	5	A	21	D	37	D
6	B	22	B			6	B	22	B	38	D
7	B	23	C			7	B	23	A	39	C
8	A	24	A			8	D	24	D	40	D
9	B	25	B			9	C	25	D	41	D
10	A	26	A			10	B	26	D	42	D
11	D	27	C			11	C	27	C	43	D
12	B	28	A			12	C	28	A	44	D
13	A	29	A			13	A	29	C	45	A
14	B	30	A			14	B	30	B	46	B
15	B	31	D			15	C	31	B	47	B
16	D	32	A			16	C	32	C		

Score Your Test

ISEE scores are broken down by four sections: Verbal Reasoning, Reading Comprehension, Quantitative Reasoning, and Mathematics Achievement. A sum of the ALL sections is also reported. The Essay section is scored separately.

For the Upper Level ISEE, the score range is 760 to 940, the lowest possible score a student can earn is 760 and the highest score is 940 for each section. A student receives 1 point for every correct answer. There is no penalty for wrong or skipped questions.

The total scaled score for an Upper Level ISEE test is the sum of the scores for all sections. A student will also receive a percentile score of between 1-99% that compares that student's test scores with those of other test takers of same grade and gender from the past 3 years. When a student receives her/his score, the percentile score is also be broken down into a stanine and the stanines are ranging from 1–9. Most schools accept students with scores of 5–9. The ideal candidate has scores of 6 or higher.

Percentile Rank	Stanine
1 – 3	1
4 – 10	2
11- 22	3
23 - 39	4
40 – 59	5
60 – 76	6
77- 88	7
89 – 95	8
96 - 99	9

The following charts provide an estimate of students ISEE percentile rankings for the practice tests, compared against other students taking these tests. Keep in mind that these percentiles are estimates only, and your actual ISEE percentile will depend on the specific group of students taking the exam in your year.

ISEE Upper Level Quantitative Reasoning Percentiles			
Grade Applying to	25th Percentile	50th Percentile	75th Percentile
9th	850	880	895
10th	855	885	900
11th	860	890	905
12th	864	892	908

Use the next table to convert ISEE Upper level raw score to scaled score for application to 9th - 12th grade.

ISEE Upper Level Scaled Scores									
Raw Score	Quantitative Reasoning		Mathematics Achievement		Raw Score	Quantitative Reasoning		Mathematics Achievement	
	Report Range		Report Range			Report Range		Report Range	
0	760	760	760	760	26	900	885	885	865
1	770	765	770	765	27	905	890	885	865
2	780	770	780	770	28	910	895	890	870
3	790	775	790	775	29	910	900	890	870
4	800	780	800	780	30	915	905	895	875
5	810	785	810	785	31	920	910	895	875
6	820	790	820	790	32	925	915	900	880
7	825	795	825	795	33	930	920	900	880
8	830	800	830	800	34	930	925	905	885
9	835	805	835	805	35	935	930	905	885
10	840	810	840	810	36	935	935	910	890
11	845	815	845	815	37	940	940	910	890
12	850	820	850	820	38			915	895
13	855	825	855	825	39			920	900
14	860	830	855	830	40			925	905
15	865	835	860	835	41			925	910
16	870	840	860	840	42			930	915
17	875	845	865	840	43			930	920
18	880	845	865	845	44			935	925
19	880	850	870	845	45			935	930
20	885	855	870	850	46			940	935
21	885	860	875	850	47			940	940
22	890	865	875	855					
23	890	870	875	855					
24	895	875	880	860					
25	895	880	880	860					

ISEE UPPER LEVEL Math Practice Test

Section 1

Answers and Explanations

1) Choice D is correct

$2y + 6 < 30 \rightarrow 2y < 30 - 6 \rightarrow 2y < 24 \rightarrow y < 12$
Only choice D (8) is less than 12.

2) Choice C is correct

A factor must divide evenly into its multiple. 16 cannot be a factor of 90 because 90 divided by $16 = 5.62$

3) Choice B is correct

The diagonal of the square is 6. Let x be the side.
Use Pythagorean Theorem: $a^2 + b^2 = c^2$
$x^2 + x^2 = 6^2 \Rightarrow 2x^2 = 6^2 \Rightarrow 2x^2 = 36 \Rightarrow x^2 = 18 \Rightarrow x = \sqrt{18}$
The area of the square is: $\sqrt{18} \times \sqrt{18} = 18$

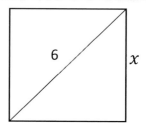

4) Choice B is correct

To find what percent A is of B, divide A by B, then multiply that number by 100%:
$13.26 \div 58.56 = 0.2264 \times 100\% = 22.64\%$,   This is approximately 22%.

5) Choice C is correct

$Probability = \frac{number\ of\ desired\ outcomes}{number\ of\ total\ outcomes}$
In this case, a desired outcome is selecting either a red or a yellow marble. Combine the number of red and yellow marbles: $9 + 6 = 15$, and divide this by the total number of marbles:
$7 + 9 + 6 = 22$. The probability is $\frac{15}{22}$.

6) Choice B is correct

Begin by calculating James's total earnings after 20 hours:
$20\ hours \times \$8.500\ per\ hour = \$170$
Next, divide this total by Jacob's hourly rate to find the number of hours Jacob would need to work: $\$170 \div \$10.00\ per\ hour = 17\ hours$

7) Choice C is correct

The total cost of the phone call can be represented by the equation: $TC = \$5.00 + \$0.5x$, where x is the duration of the call after the first five minutes. In this case, $x = 30$. Substitute the known values into the equation and solve: $TC = \$5.00 + \$0.5 \times 30$
$TC = \$5.00 + \$15.00 , \qquad TC = \$20.00$

8) Choice B is correct

Let b be the amount of time Alec can do the job, then,
$$\frac{1}{a} + \frac{1}{b} = \frac{1}{50} \rightarrow \frac{1}{150} + \frac{1}{b} = \frac{1}{50} \rightarrow \frac{1}{b} = \frac{1}{50} - \frac{1}{150} = \frac{2}{150} = \frac{1}{75}$$
Then: $b = 75$ minutes

9) Choice A is correct

First, find the number. Let x be the number. Write the equation and solve for x.
150% of a number is 75, then: $1.5 \times x = 75 \rightarrow x = 75 \div 1.5 = 50$
90% of 50 is: $0.9 \times 50 = 45$

10) Choice B is correct

The length of MN is equal to:$4x + 6x = 10x$, Then: $10x = 50 \rightarrow x = \frac{50}{10} = 5$
The length of ON is equal to: $6x = 6 \times 5 = 30 \ cm$

11)   Choice A is correct
The distance on the map is proportional to the actual distance between the two cities. Use the information to set up a proportion and then solve for the unknown number of actual miles:
$\frac{14 \ miles}{\frac{1}{2} \ inches} = \frac{x \ miles}{18 \ inches}$,   Cross multiply and simplify to solve for the x:

$\frac{14 \times 18}{\frac{1}{2}} = x \ miles \rightarrow \frac{252}{\frac{1}{2}} = 252 \times 2 = 504 \ miles$

12)   Choice B is correct
Number of Mathematics book: $0.35 \times 840 = 294$
Number of English books: $0.15 \times 840 = 126$
Product of number of Mathematics and number of English books: $294 \times 126 = 37{,}044$

13) Choice D is correct

If each book weighs $\frac{1}{6}$ pound, then $1 \ pound = 6 \ books$. To find the number of books in 60 pounds, simply multiply this 6 by 60: $60 \times 6 = 360$

14) Choice B is correct

Write equations based on the information provided in the question:
$Emily = Daniel, Emily = 6 \ Claire, Daniel = 15 + \ Claire$
$Emily = Daniel \ \rightarrow Emily = 15 + Claire$
$Emily = 6 \ Claire \ \rightarrow 6 \ Claire = 15 + Claire \rightarrow 6 \ Claire - \ Claire = 15$

$5\,Claire = 15, Claire = 3$

15) Choice A is correct

Begin by examining the sequence to find the pattern. The difference between 3 and 6 is 3; moving from 6 to 11 requires 5 to be added; moving from 11 to 18 requires 7 to be added. The pattern emerges here — adding by consecutive odd integers. The 5th term is equal to $18 + 9 = 27$, and the 6th term is equal to $27 + 11 = 38$.

16) Choice B is correct

The general slope-intercept form of the equation of a line is $y = mx + b$, where m is the slope and b is the y-intercept. By substitution of the given point and given slope: $-3 = (3)(7) + b$, So, $b = -3 - 21 = -24$, and the required equation is $y = 3x - 24$.

17) Choice D is correct

Multiplying each side of $-4x - y = 8$ by 2 gives $-8x - 2y = 16$. Adding each side of $-8x - 2y = 16$ to the corresponding side of $8x + 6y = 20$ gives $4y = 36$ or $y = 9$. Finally, substituting 9 for y in $8x + 6y = 20$ gives $8x + 6(9) = 20$ or $x = -\frac{17}{4}$.

18) Choice B is correct

The question is this: 530.20 is what percent of 631? Use percent formula:
$Part = \frac{percent}{100} \times whole,$
$530.20 = \frac{percent}{100} \times 631 \rightarrow 530.20 = \frac{percent \times 631}{100} \rightarrow 53020 = percent \times 631$
Then, Percent $= \frac{53020}{631} = 84.02$
530.20 is 84% of 631. Therefore, the discount is: $100\% - 84\% = 16\%$

19) Choice A is correct

Let x be the number. Write the equation and solve for x. $\frac{3}{2} \times 20 = \frac{5}{2}x \rightarrow \frac{3 \times 20}{2} = \frac{5x}{2}$, use cross multiplication to solve for x. $2 \times 60 = 5x \times 2 \Rightarrow 120 = 10x \Rightarrow x = 12$

20) Choice A is correct

Let's choose $100 for the sales of the supermarket. If the sales increases by 11 percent in April, the final amount of sales at the end of April will be $100 + (11\%) \times (\$100) = \$111$.
If sales then decreased by 11 percent in May, the final amount of sales at the end of August will be $\$111 - (\$111) \times (11\%) = \$98.79$
The final sales of $98 is 98% of the original price of $100. Therefore, the sales decreased by 2% overall.

21) Choice B is correct

The time it takes to drive from city A to city B is: $\frac{2700}{74} = 36.48$

22) Choice C is correct

Recall that the first and only even prime number is 2. The other prime numbers are: $3, 5, 7, 9, 11, 13, 17$, etc. They are all odd numbers except for 2, so the sum of members in Set A is just 2. So, the correct answer is C.

23) Choice B is correct

Column A: $5\% = \frac{5}{100} = 0.05$,   Column B: $\frac{1}{2} = 0.5$,   0.5 is greater than 0.05

24) Choice B is correct

Column A: $\frac{12 + 18 + 26 + 30 + 32}{5} = \frac{118}{5} = 23.6$

Column B: $\frac{24 + 28 + 30 + 36}{4} = \frac{118}{4} = 29.5$

25) Choice A is correct

The posts are placed 12.5 feet apart. Since a post is needed at the very beginning as well as at the end, A requires 9 posts. $100 \div 12.5 = 8$, $8 + 1 = 9$

26) Choice D is correct

First, solve the expression for x. $\frac{x}{4} = y^2 \rightarrow x = 4y^2$

Plug in different values for y and find the values of x.

Let's choose $y = 0 \rightarrow x = 4y^2 \rightarrow x = 4(0)^2 = 0$

The values in Column A and B are equal.

Now, let's choose $y = 1 \rightarrow x = 4y^2 \rightarrow x = 4(1)^2 = 4$

Column A is greater. So, the relationship cannot be determined from the information given.

27) Choice A is correct

Column A: $3x^2 - 2x + 4 = 3(-1)^2 - 2(-1) + 4 = 3 + 2 + 4 = 9$

Column B: $2x^3 + x^2 + 4 = 2(-1)^3 + (-1)^2 + 4 = -2 + 1 + 4 = 3$

28) Choice A is correct

Column A: $9 + 12(9 - 5) = 57$

Column B: $12 + 9(9 - 5) = 48$

29) Choice C is correct

First find the value of x. $\frac{x}{48} = \frac{2}{3} \rightarrow 3x = 2 \times 48 = 96 \rightarrow x = \frac{96}{3} = 32$

Column A: $\frac{8}{x} = \frac{8}{32} = \frac{1}{4}$

30) Choice A is correct

First convert hours to minutes. 2 hours 5 minutes $= 2 \times 60 + 5 = 125$ minutes.

Machine D makes b rolls of steel in 25 minutes. So, it makes 5 sets of b in 125 minutes.
$125 \div 25 = 5$ sets of b.
Machine E operates for 4 hours, making b rolls per hour. So, it makes a total of $4b$ rolls.
Therefore, machine D makes more rolls, and Column A is greater.

31) Choice D is correct

$6 > y > -2,$ Let's choose some values for y. $y = -1$
Column A: $\frac{y}{4} = \frac{-1}{4}$
Column B: $\frac{4}{y} = \frac{4}{-1} = -1$
In this case, column A is bigger.
$y = 1$
Column A: $\frac{y}{4} = \frac{1}{4}$
Column B: $\frac{4}{y} = \frac{4}{1} = 4$
In this case, Column B is bigger. So, the relationship cannot be determined from the information given.

32) Choice D is correct

$\frac{a}{b} = \frac{c}{d}$, Here there are two equal fractions. Let's choose some values for these variables. $\frac{1}{2} = \frac{2}{4}$
In this case, Column A is $3 (1 + 2)$ and Column B is $6 (2 + 4)$. Since, we can change the positions of these variables (for example put 2 for a and 4 for b), he relationship cannot be determined from the information given.

33) Choice A is correct

The ratio of boys to girls in a class is 7 to 11. Therefore, ratio of boys to the entire class is 7 out of 18. $\frac{7}{18} > \frac{1}{3}$

34) Choice A is correct

Let us calculate each probability individually:
That probability that the first marble is blue $= \frac{6}{10} = \frac{3}{5}$
The probability that the second marble is blue $= \frac{5}{9}$
Column A: The probability that both marbles are blue $= \frac{3}{5} \times \frac{5}{9} = \frac{15}{45} = \frac{1}{3}$
The probability that the marble is green $= \frac{4}{10} = \frac{2}{5}$
The probability that the second marble is blue $= \frac{6}{9} = \frac{2}{3}$
Column B: The probability that the first marbles is green, but the second is blue $= \frac{2}{5} \times \frac{2}{3} = \frac{4}{15}$
Column A is greater. $\frac{1}{3} > \frac{4}{15}$.

35) Choice A is correct

First, let's find the number of digits when the printer prints 100 pages.
If there are 2 digits in each page and the printer prints 100 pages, then, there will be 200 digits.
$100 \times 2 = 200$
However, we know that pages $1 - 9$ have only one digit each, so we must subtract 9 from this total: $200 - 9 = 191$. We also know that the number 100th has three digits not two. So, we must add 1 digit to this total: $191 + 1 = 192$.
It is given that 195 digits were printed, and we know that 100 pages results in 192 digits total, so there must be 101 total pages in the magazine. Column A is greater.

36) Choice B is correct

Column A: The largest number that can be written by rearranging the digits in $263 = 632$
Column B: The largest number that can be written by rearranging the digits in $192 = 921$

37) Choice B is correct

The computer priced \$124 includes 4% profit. Let x be the original cost of the computer.
Then: $x + 4\% \ of \ x = 124 \rightarrow x + 0.04x = 124 \rightarrow 1.04x = 124 \rightarrow x = \frac{124}{1.04} = 119.23$
Column B is bigger.

ISEE UPPER LEVEL Math Practice Test

Section 2

Answers and Explanations

1)   Choice A is correct.

$0.00003379 = \frac{3.379}{100,000} \Rightarrow 3.379 \times 10^{-5}$

2)   Choice A is correct

$|10 - (12 \div |1 - 5|)| = |10 - (12 \div |-4|)| = |10 - (12 \div 4)| = |10 - 3| = |7| = 7$

3) Choice A is correct

Use FOIL (First, Out, In, Last) method.
$(x + 4)(x + 5) = x^2 + 5x + 4x + 20 = x^2 + 9x + 20$

4)   Choice B is correct

Use quadratic formula: $ax^2 + bx + c = 0$
$x_{1,2} = \frac{-b \pm \sqrt{b^2 - 4ac}}{2a}$
$6x^2 + 16x + 8 \quad \Rightarrow \qquad$ then: $a = 6, b = 16$ and $c = 8$

$$x = \frac{-16 + \sqrt{16^2 - 4 \times 6 \times 8}}{2 \times 6} = -\frac{2}{3}$$

$$x = \frac{-16 - \sqrt{16^2 - 4 \times 6 \times 8}}{2 \times 6} = -2$$

5) Choice D is correct

Solve for x. $-2 \leq 2x - 4 < 8 \Rightarrow$ (add 4 all sides) $-2 + 4 \leq 2x - 4 + 4 < 8 + 4 \Rightarrow$
$2 \leq 2x < 12 \Rightarrow$ (divide all sides by 2) $1 \leq x < 6$
x is between 1 and 6. Choice D represent this inequality.

6)   Choice C is correct

Simplify: $1 - 9 \div (4^2 \div 2) = 1 - 9 \div (16 \div 2) = -8 \div 8 = -1$

7)   Choice D is correct.

Write the proportion and solve for missing side.
$$\frac{\text{Smaller triangle height}}{\text{Smaller triangle base}} = \frac{\text{Bigger triangle height}}{\text{Bigger triangle base}} \Rightarrow \frac{80cm}{200cm} = \frac{80 + 460cm}{x} \Rightarrow x = 1350 \ cm$$

8)   Choice B is correct

The ratio of boy to girls is $3 : 5$. Therefore, there are 3 boys out of 8 students. To find the answer, first divide the total number of students by 8, then multiply the result by 3.
$600 \div 8 = 75 \Rightarrow 75 \times 3 = 225$

9)   Choice B is correct.

A linear equation is a relationship between two variables, x and y, and can be written in the form of $y = mx + b$
A non-proportional linear relationship takes on the form $y = mx + b$, where $b \neq 0$ and its graph is a line that does not cross through the origin.
Only in graph B, the line does not pass through the origin.

10) Choice D is correct

$90 \div \frac{1}{9} = 90 \times 9 = 810$

11) Choice B is correct.

Translated 5 units down and 4 units to the left means: $(x . y) \Rightarrow (x - 4, y - 5)$

12) Choice A is correct

Area of a rectangle = $width \times height$
$Area = 148 \times 90 = 13320$

13) Choice C is correct.

$\frac{4}{23} = 0.173$ and $25\% = 0.25$ therefore x should be between 0.173 and 0.25

E. Only choice B $(\sqrt{0.044}) = 0.20$ is between 0.173 and 0.25

14) Choice C is correct

All Integers must end in one of the following digits:
0 when multiplied by itself ends in 0
1 when multiplied by itself ends in 1
2 when multiplied by itself ends in 4
3 when multiplied by itself ends in 9
4 when multiplied by itself ends in 6
5 when multiplied by itself ends in 5
6 when multiplied by itself ends in 6
7 when multiplied by itself ends in 9
8 when multiplied by itself ends in 4
9 when multiplied by itself ends in 1
Number 8 is not in the results.

15) Choice C is correct

The area of the trapezoid is: $Area = \frac{1}{2}h(b_1 + b_2) = \frac{1}{2}(x)(13 + 8) = 126$
$\rightarrow 10.5x = 126 \rightarrow x = 12$
$y = \sqrt{5^2 + 12^2} = \sqrt{25 + 144} = \sqrt{169} = 13$
The perimeter of the trapezoid is: $12 + 13 + 8 + 13 = 46$

16) Choice A is correct

$\frac{1}{2}$ of the distance $4\frac{1}{5}$ miles is: $\frac{1}{2} \times 4\frac{1}{5} = \frac{1}{2} \times \frac{21}{5} = \frac{21}{10}$
Converting $\frac{21}{10}$ to a mixed number gives: $\frac{21}{10} = 2\frac{1}{10}$

17) Choice B is correct

$average = \frac{sum}{total} = \frac{40 + 45 + 50}{3} = \frac{135}{3} = 45$

18) Choice B is correct

Use Pythagorean theorem: $a^2 + b^2 = c^2 \rightarrow s^2 + h^2 = (3s)^2 \rightarrow s^2 + h^2 = 9s^2$
Subtracting s^2 from both sides gives: $h^2 = 8s^2$
Square roots of both sides: $h = \sqrt{8s^2} = \sqrt{4 \times 2 \times s^2} = \sqrt{4} \times \sqrt{2} \times \sqrt{s^2} = 2 \times s \times \sqrt{2} = 2s\sqrt{2}$

19) Choice C is correct

Let x be the number of purple marbles. Let's review the choices provided:

A. $\frac{1}{11}$, if the probability of choosing a purple marble is one out of ten, then:

$Probability = \frac{number\ of\ desired\ outcomes}{number\ of\ total\ outcomes} = \frac{x}{40+40+30+x} = \frac{1}{11}$

Use cross multiplication and solve for x. $11x = 110 + x \rightarrow 10x = 110 \rightarrow x = 11$

Since, number of purple marbles can be 9, then, choice be the probability of randomly selecting a purple marble from the bag.

Use same method for other choices.

B. $\frac{1}{6}$

$\frac{x}{40+40+30+x} = \frac{1}{6} \rightarrow 6x = 110 + x \rightarrow 5x = 110 \rightarrow x = 22$

C. $\frac{2}{5}$

$\frac{x}{40+40+30+x} = \frac{2}{5} \rightarrow 5x = 220 + 2x \rightarrow 3x = 220 \rightarrow x = 73.3$

D. $\frac{1}{23}$

$\frac{x}{40+40+30+x} = \frac{1}{23} \rightarrow 23x = 110 + x \rightarrow 22x = 110 \rightarrow x = 5$

Number of purple marbles cannot be a decimal.

20) Choice C is correct

Let x be the original price of the dress. Then: 23% of $x = 21.87$

$x = \frac{23}{100}x = 21.87$

$x = \frac{100 \times 21.87}{23} \cong 95.08$

21) Choice D is correct

$\frac{8}{35} = 0.22$

22) Choice B is correct

30% of A is 1,200 Then: $0.3A = 1,200 \rightarrow A = \frac{1,200}{0.3} = 4,000$

12% of 4,000 is: $0.12 \times 4,000 = 480$

23) Choice D is correct

Simplify:

$$\frac{\frac{1}{2} - \frac{x+5}{4}}{\frac{x^3}{2} - \frac{5}{2}} = \frac{\frac{1}{2} - \frac{x+5}{4}}{\frac{x^3 - 5}{2}} = \frac{2(\frac{1}{2} - \frac{x+5}{4})}{x^3 - 5}$$

$\Rightarrow$ Simplify: $\frac{1}{2} - \frac{x+5}{4} = \frac{-x-3}{4}$

Then: $\dfrac{2\left(\frac{-x-3}{4}\right)}{x^3-5} = \dfrac{\frac{-x-3}{2}}{x^3-5} = \dfrac{-x-3}{2(x^3-5)} = \dfrac{-x-3}{2x^3-10}$

24) Choice A is correct

$(6.2 + 8.3 + 2.4) \times x = x$, $16.9x = x$, Then: $x = 0$

25) Choice D is correct

For sum of 5: $(1\ \&\ 4)$ *and* $(4\ \&\ 1)$, $(2\ \&\ 3)$ and $(3\ \&\ 2)$, therefore we have 4 options.
For sum of 8: $(5\ \&\ 3)$ *and* $(3\ \&\ 5)$, $(4\ \&\ 4)$ and $(2\ \&\ 6)$, $(6\ \&\ 2)$, $(1\ \&7)$ and $(7\ \&\ 1)$ we have 7 options. To get a sum of 5 or 8 for two dice: $4 + 7 = 11$
Since, we have $6 \times 6 = 36$ total number of options, the probability of getting a sum of 5 and 8 is 11 out of 36 or $\dfrac{11}{36}$

26) Choice C is correct

Write a proportion and solve. $\dfrac{7}{38} = \dfrac{x}{190} \rightarrow x = \dfrac{7 \times 190}{38} = 35$

27) Choice B is correct

$78.56 \div 0.05 = 1571.2$

28) Choice D is correct

The distance of A to B on the coordinate plane is: $\sqrt{(x_1 - x_2)^2 + (y_1 - y_2)^2} =$
$\sqrt{(2 - 10)^2 + (6 - 12)^2} = \sqrt{8^2 + 6^2}$
$= \sqrt{64 + 36} = \sqrt{100} = 10$
The diameter of the circle is 10 and the radius of the circle is 5. Then: the circumference of the circle is: $2\pi r = 2\pi(5) = 10\pi$

29) Choice C is correct

$2x^2 + 6 = 26 \rightarrow 2x^2 = 20 \rightarrow x^2 = 10 \rightarrow x = \pm\sqrt{10}$

30) Choice A is correct

Diameter = 20, then: Radius = 10
Area of a circle $= \pi r^2 \qquad \Rightarrow A = 3.14(10)^2 = 314$

31) Choice A is correct

$6\ days\ 20\ hours\ 36\ minutes - 4\ days\ 12\ hours\ 24\ minutes =$
$2\ days\ 8\ hours\ and\ 12 minutes$

32) Choice C is correct

Formula of triangle area $= \dfrac{1}{2}(base \times height)$

Since the angles are $45 - 45 - 90$, then this is an isosceles triangle, meaning that the base and height of the triangle are equal.

$$Triangle\ area\ =\ \frac{1}{2}\ (base \times height) = \frac{1}{2}(4 \times 4) = 8$$

33) Choice B is correct

The mean of the data is 11. Then:
$$\frac{x+10+15+13+6}{5} = 11 \rightarrow x + 44 = 55 \rightarrow x = 55 - 44 = 11$$

34) Choice C is correct

The difference of the file added, and the file deleted is:
$542,159 - 499,986 = 42,173$
$837,036 + 42,173 = 879,209$

35) Choice B is correct

$40\%\ of\ 70 = 28 \rightarrow 70 + 28 = 98$

36) Choice D is correct

$x^{\frac{1}{2}}$equals to the root of x. Then: $12 + x^{\frac{1}{2}} = 24 \rightarrow 12 + \sqrt{x} = 24 \rightarrow \sqrt{x} = 12 \rightarrow x = 144$
$x = 144$ and $5 \times x$ equals: $5 \times 144 = 720$

37) Choice C is correct

Only choice C represents the statement "twice the difference between 5 times H and 2 gives 35". $2(5H - 2) = 35$

38) Choice D is correct

The area of the circle is $25\pi\ cm^2$, then, its diameter is $10cm$.
$area\ of\ a\ circle = \pi r^2 = 25\pi \rightarrow r^2 = 25 \rightarrow r = 5$
Radius of the circle is 5 and diameter is twice of it, 10.
One side of the square equals to the diameter of the circle. Then:
$Area\ of\ square = side \times side = 10 \times 10 = 100$

39) Choice C is correct.

When a point is reflected over x axes, the (y) coordinate of that point changes to $(-y)$ while its x coordinate remains the same. C $(9, 7) \rightarrow$ C' $(9, -7)$

40) Choice D is correct.

A set of ordered pairs represents y as a function of x if: $x_1 = x_2 \rightarrow y_1 = y_2$
In choice A: $(5, -2)$ and $(5, 7)$ are ordered pairs with same x and different y, therefore y isn't a function of x.

In choice B: $(2, 2)$ and $(2, 7)$ are ordered pairs with same x and different y, therefore y isn't a function of x.
In choice C: $(8, 7)$ and $(8, 18)$ are ordered pairs with same x and different y, therefore y isn't a function of x.

41) Choice D is correct

If the length of the box is 24, then the width of the box is one third of it, 8, and the height of the box is 4 (half of the width). The volume of the box is:
$V = length \times width \times height = (24)\,(8)\,(4) = 768$

42) Choice D is correct

Number of squares equal to: $\frac{40 \times 12}{4 \times 4} = 10 \times 3 = 30$

43) Choice A is correct

David's weekly salary is $230 plus 9% of $1,200. Then: $9\% \ of \ 1,200 = 0.09 \times 1,200 = 108$
$230 + 108 = 338$

44) Choice B is correct

$5x^3y^2 + 4x^5y^3 - (6x^3y^2 - 4x^1y^5) = 5x^3y^2 + 4x^5y^3 - 6x^3y^2 + 4x^1y^5)$
$$= 4x^5y^3 - x^3y^2 + 4xy^5$$

45) Choice B is correct

Let P be circumference of circle A, then; $2\pi r_A = 20\pi \rightarrow r_A = 10$
$r_A = 5r_B \rightarrow r_B = \frac{10}{5} = 2 \rightarrow$ Area of circle B is; $\quad \pi r_B^2 = 4\pi$

46) Choice D is correct

The area of the square is 64 square inches. $Area \ of \ square = side \times side = 8 \times 8 = 64$
The length of the square is increased by 5 inches and its width decreased by 4 inches. Then, its area equals: $Area \ of \ rectangle = width \times length = 13 \times 4 = 52$
The area of the square will be decreased by 12 square inches. $64 - 52 = 12$

47) Choice D is correct

Write a proportion and solve. $\frac{3}{2} = \frac{x}{80}$
Use cross multiplication: $2x = 240 \rightarrow x = 120$

Day 30:
A Realistic ISEE Upper Level Math Test

Time to experience a REAL ISEE Upper Level Math Test

Take the following practice ISEE Upper Level Math Test to simulate the test day experience. After you've finished, score your test using the answer key.

Before You Start

- Keep your practice test experience as realistic as possible.

- You'll need a pencil and a timer to take the test.

- After you've finished the test, review the answer key to see where you went wrong.

- **Keep Strict Timing on the Test Section!**

Good Luck!

Mathematics is not only real, but it is the only reality. ~ Martin Gardner

ISEE Upper Level Math

Practice Test 2019

Two Parts

Total number of questions: 84

Part 1 (Quantitative Reasoning): 37 questions

Part 2 (Mathematics Achievement): 47 questions

Total time for two parts: 75 Minutes

ISEE Upper Level Practice Test
Quantitative Reasoning

1) Ⓐ Ⓑ Ⓒ Ⓓ   2) Ⓐ Ⓑ Ⓒ Ⓓ

3) Ⓐ Ⓑ Ⓒ Ⓓ   4) Ⓐ Ⓑ Ⓒ Ⓓ

5) Ⓐ Ⓑ Ⓒ Ⓓ   6) Ⓐ Ⓑ Ⓒ Ⓓ

7) Ⓐ Ⓑ Ⓒ Ⓓ   8) Ⓐ Ⓑ Ⓒ Ⓓ

9) Ⓐ Ⓑ Ⓒ Ⓓ   10) Ⓐ Ⓑ Ⓒ Ⓓ

11) Ⓐ Ⓑ Ⓒ Ⓓ   12) Ⓐ Ⓑ Ⓒ Ⓓ

13) Ⓐ Ⓑ Ⓒ Ⓓ   14) Ⓐ Ⓑ Ⓒ Ⓓ

15) Ⓐ Ⓑ Ⓒ Ⓓ   16) Ⓐ Ⓑ Ⓒ Ⓓ

17) Ⓐ Ⓑ Ⓒ Ⓓ   18) Ⓐ Ⓑ Ⓒ Ⓓ

19) Ⓐ Ⓑ Ⓒ Ⓓ   20) Ⓐ Ⓑ Ⓒ Ⓓ

21) Ⓐ Ⓑ Ⓒ Ⓓ   22) Ⓐ Ⓑ Ⓒ Ⓓ

23) Ⓐ Ⓑ Ⓒ Ⓓ   24) Ⓐ Ⓑ Ⓒ Ⓓ

25) Ⓐ Ⓑ Ⓒ Ⓓ   26) Ⓐ Ⓑ Ⓒ Ⓓ

27) Ⓐ Ⓑ Ⓒ Ⓓ   28) Ⓐ Ⓑ Ⓒ Ⓓ

29) Ⓐ Ⓑ Ⓒ Ⓓ   30) Ⓐ Ⓑ Ⓒ Ⓓ

31) Ⓐ Ⓑ Ⓒ Ⓓ   32) Ⓐ Ⓑ Ⓒ Ⓓ

33) Ⓐ Ⓑ Ⓒ Ⓓ   34) Ⓐ Ⓑ Ⓒ Ⓓ

35) Ⓐ Ⓑ Ⓒ Ⓓ   36) Ⓐ Ⓑ Ⓒ Ⓓ

37) Ⓐ Ⓑ Ⓒ Ⓓ

ISEE Upper Level Practice Test
Mathematics Achievement

1) Ⓐ Ⓑ Ⓒ Ⓓ   2) Ⓐ Ⓑ Ⓒ Ⓓ

3) Ⓐ Ⓑ Ⓒ Ⓓ   4) Ⓐ Ⓑ Ⓒ Ⓓ

5) Ⓐ Ⓑ Ⓒ Ⓓ   6) Ⓐ Ⓑ Ⓒ Ⓓ

7) Ⓐ Ⓑ Ⓒ Ⓓ   8) Ⓐ Ⓑ Ⓒ Ⓓ

9) Ⓐ Ⓑ Ⓒ Ⓓ   10) Ⓐ Ⓑ Ⓒ Ⓓ

11) Ⓐ Ⓑ Ⓒ Ⓓ   12) Ⓐ Ⓑ Ⓒ Ⓓ

13) Ⓐ Ⓑ Ⓒ Ⓓ   14) Ⓐ Ⓑ Ⓒ Ⓓ

15) Ⓐ Ⓑ Ⓒ Ⓓ   16) Ⓐ Ⓑ Ⓒ Ⓓ

17) Ⓐ Ⓑ Ⓒ Ⓓ   18) Ⓐ Ⓑ Ⓒ Ⓓ

19) Ⓐ Ⓑ Ⓒ Ⓓ   20) Ⓐ Ⓑ Ⓒ Ⓓ

21) Ⓐ Ⓑ Ⓒ Ⓓ   22) Ⓐ Ⓑ Ⓒ Ⓓ

23) Ⓐ Ⓑ Ⓒ Ⓓ   24) Ⓐ Ⓑ Ⓒ Ⓓ

25) Ⓐ Ⓑ Ⓒ Ⓓ   26) Ⓐ Ⓑ Ⓒ Ⓓ

27) Ⓐ Ⓑ Ⓒ Ⓓ   28) Ⓐ Ⓑ Ⓒ Ⓓ

29) Ⓐ Ⓑ Ⓒ Ⓓ   30) Ⓐ Ⓑ Ⓒ Ⓓ

31) Ⓐ Ⓑ Ⓒ Ⓓ   32) Ⓐ Ⓑ Ⓒ Ⓓ

33) Ⓐ Ⓑ Ⓒ Ⓓ   34) Ⓐ Ⓑ Ⓒ Ⓓ

35) Ⓐ Ⓑ Ⓒ Ⓓ   36) Ⓐ Ⓑ Ⓒ Ⓓ

37) Ⓐ Ⓑ Ⓒ Ⓓ   38) Ⓐ Ⓑ Ⓒ Ⓓ

39) Ⓐ Ⓑ Ⓒ Ⓓ   40) Ⓐ Ⓑ Ⓒ Ⓓ

41) Ⓐ Ⓑ Ⓒ Ⓓ   42) Ⓐ Ⓑ Ⓒ Ⓓ

43) Ⓐ Ⓑ Ⓒ Ⓓ   44) Ⓐ Ⓑ Ⓒ Ⓓ

45) Ⓐ Ⓑ Ⓒ Ⓓ   46) Ⓐ Ⓑ Ⓒ Ⓓ

47) Ⓐ Ⓑ Ⓒ Ⓓ

ISEE Upper Level Math Practice Test 2019

Section 1

Quantitative Reasoning

37 questions

Total time for this section: 35 Minutes

You may NOT use a calculator for this test.

1) How much greater is the value of $4x + 9$ than the value of $4x - 3$?

A. 8

B. 10

C. 12

D. 14

2) What is the prime factorization of 1400?

A. $2 \times 2 \times 5 \times 5$

B. $2 \times 2 \times 2 \times 5 \times 5 \times 7$

C. 2×5

D. $2 \times 2 \times 2 \times 5 \times 7$

3) If 6 inches on a map represents an actual distance of 150 feet, then what actual distance does 20 inches on the map represent?

A. 180

B. 200

C. 250

D. 500

4) The circle graph below shows all Mr. Green's expenses for last month. If he spent $550 on his car, how much did he spend for his rent?

A. $675

B. $750

C. $780

D. $810

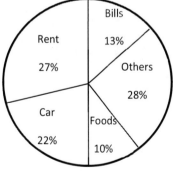

Mr. Green's monthly expenses

5) The area of a circle is less than 49π. Which of the following can be the circumference of the circle?

A. 10π

B. 14π

C. 24π

D. 32π

6) A basket contains 25 balls and the average weight of each of these balls is 35 g. The five heaviest balls have an average weight of 50 g each. If we remove the three heaviest balls from the basket, what is the average weight of the remaining balls?

A. 10

B. 20.25

C. 31.25

D. 35

7) If $f(x) = x^2 + 6$, what is the smallest possible value of $f(x)$?

A. 0

B. 5

C. 6

D. 7

8) Alice drives from her house to work at an average speed of 45 miles per hour and she drives at an average speed of 65 miles per hour when she was returning home. What was her minimum speed on the round trip in miles per hour?

A. 45

B. 58.5

C. 65

D. Cannot be determined

9) If the sum of the positive integers from 1 to n is 3,350, and the sum of the positive integers from $n + 1$ to $2n$ is 4,866, which of the following represents the sum of the positive integers from 1 to $2n$ inclusive?

A. 3,350

B. 4,866

C. 7,000

D. 8216

10) Oscar purchased a new hat that was on sale for \$8.34. The original price was \$14.65. What percentage discount was the sale price?

A. 4.2%

B. 40.5%

C. 43%

D. 45%

11) Which of the following statements is correct, according to the graph below?

A. Number of books sold in April was twice the number of books sold in July.

B. Number of books sold in July was less than half the number of books sold in May.

C. Number of books sold in June was half the number of books sold in April.

D. Number of books sold in July was equal to the number of books sold in April plus the number of books sold in June.

12) List A consists of the numbers $\{2, 4, 9, 11, 16\}$, and list B consists of the numbers $\{5, 7, 13, 15, 18\}$.

   If the two lists are combined, what is the median of the combined list?

A. 7

B. 8

C. 9

D. 10

13) A bag contains 19 balls: three green, five black, eight blue, a brown, a red and one white. If 17 balls are removed from the bag at random, what is the probability that a brown ball has been removed?

A. $\dfrac{1}{9}$

B. $\dfrac{1}{6}$

C. $\dfrac{16}{19}$

D. $\dfrac{17}{19}$

14) If Jim adds 100 stamps to his current stamp collection, the total number of stamps will be equal to $\dfrac{5}{6}$ the current number of stamps. If Jim adds 50% more stamps to the current collection, how many stamps will be in the collection?

A. 150

B. 300

C. 600

D. 900

15) If $x + y = 7$ and $x - y = 6$ then what is the value of $(x^2 - y^2)$?

A. 24

B. 42

C. 65

D. 90

16) The area of rectangle $ABCD$ is 100 square inches. If the length of the rectangle is three times the width, what is the perimeter of rectangle $ABCD$?

A. 40

B. 67

C. 76

D. 86

17) What's The ratio of boys and girls in a class is $7:4$. If there are 55 students in the class, how many more girls should be enrolled to make the ratio $1:1$?

A. 6

B. 10

C. 12

D. 15

18) The sum of 8 numbers is greater than 320 and less than 480. Which of the following could be the average (arithmetic mean) of the numbers?

A. 30

B. 35

C. 40

D. 45

19) A gas tank can hold 35 gallons when it is $\frac{5}{2}$ full. How many gallons does it contain when it is full?

A. 125

B. 62.5

C. 50

D. 14

20) Triangle ABC is similar to triangle ADE. What is the length of side EC?

A. 4

B. 10

C. 18

D. 45

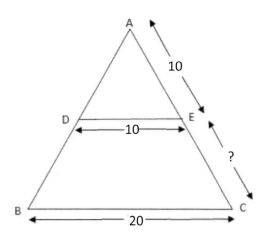

21) Which of the following expressions gives the value of b in terms of a, c, and z from the following equation?

$$a = [\frac{cz}{b}]^2$$

A. $b = ac^2z^2$

B. $b = \frac{cz}{\sqrt{a}}$

C. $b = \frac{\sqrt{a}}{cz}$

D. $b = [\frac{cz}{a}]^2$

Quantitative Comparisons

Direction: Questions 22 to 37 are Quantitative Comparisons Questions. Using the information provided in each question, compare the quantity in column A to the quantity in Column B. Choose on your answer sheet grid

    A   if the quantity in Column A is greater

    B   if the quantity in Column B is greater

    C   if the two quantities are equal

    D   if the relationship cannot be determined from the information given

22)

Column A	Column B
2^2	$\sqrt[3]{64}$

23)

Column A	Column B
8	$(59)^{\frac{1}{2}}$

24)

Column A	Column B
The average of $15, 25,$ and 27	20

25)

Column A	Column B
$11 \times 675 \times 24$	$18 \times 675 \times 17$

26) x is an integer

Column A	Column B
x	$\dfrac{x}{-2}$

27)

Column A	Column B
$(\dfrac{1}{5})^2$	5^{-2}

28) $x + 2 > 5x$

Column A	Column B
x	-1

29)

Column A	Column B
The greatest value of x in	The greatest value of x in
$6\,|2x - 4| = 6$	$6\,|2x + 4| = 6$

30) x is an integer

Column A	Column B
$(x)^3(x)^4$	$(x^3)^4$

31)

Column A	Column B
The probability that	The probability that
event x will occur.	event x will not occur.

32) The selling price of a sport jacket including 25% discount is $51.

Column A	Column B
Original price of the sport jacket	$67

33) $x^2 - 5x + 20 = 6$

Column A	Column B
x	0

34)

Column A	Column B
$(0.69)^{34}$	$(0.69)^{33}$

35)

Column A	Column B
The probability of rolling a 6 on a die and getting heads on a coin toss.	The probability of rolling an even number on a die and picking a spade from a deck of 52 cards.

36)

Column A	Column B
$0.41	Sum of one quarter, two nickels, and three pennies

37) x is an odd integer, and y is an even integer. In a certain game an odd number is considered greater than an even number.

Column A	Column B
$(x + y)^2 - y$	$(y)(x - y)$

IF YOU FINISH BEFORE TIME IS CALLED, YOU MAY CHECK YOUR WORK ON THIS SECTION ONLY. DO NOT TURN TO ANY OTHER SECTION IN THE TEST. **STOP**

ISEE Upper Level Math Practice Test 2019

Section 2

Mathematics Achievement

47 questions

Total time for this section: 40 Minutes

You may NOT use a calculator for this test.

1) Which of the following points lies on the line $4x + 6y = 20$?

A. $(2, 1)$
B. $(-1, 3)$
C. $(-2, 2)$
D. $(2, 2)$

2) 5 less than twice a positive integer is 91. What is the integer?

A. 40
B. 41
C. 42
D. 48

3) If $\dfrac{|3+x|}{5} \leq 8$, then which of the following is correct?

A. $-43 \leq x \leq 37$
B. $-43 \leq x \leq 32$
C. $-32 \leq x \leq 38$
D. $-32 \leq x \leq 32$

4) $\dfrac{1}{5b^2} + \dfrac{1}{5b} = \dfrac{1}{b^2}$, then $b = ?$

A. $-\dfrac{16}{5}$
B. 4
C. $-\dfrac{5}{16}$
D. 8

5) An angle is equal to one fifth of its supplement. What is the measure of that angle?

A. 30
B. 40
C. 60
D. 80

6) 1.3 is what percent of 26?

A. 1.3
B. 5
C. 18
D. 24

7) The cost, in thousands of dollars, of producing x thousands of textbooks is $C(x) = x^2 + 2x$. The revenue, also in thousands of dollars, is $R(x) = 40x$. find the profit or loss if 20 textbooks are produced. ($profit = revenue - cost$)
A. \$2,160 profit
B. \$360 profit
C. \$2,160 loss
D. \$360 loss

8) Simplify $7x^3y^3(2x^3y)^3 =$

A. $14x^4y^6$
B. $14x^8y^6$
C. $56x^{12}y^6$
D. $56x^8y^6$

9) Ella (E) is 5 years older than her friend Ava (A) who is 4 years younger than her sister Sofia (S). If E, A and S denote their ages, which one of the following represents the given information?

A. $\begin{cases} E = A + 5 \\ S = A - 4 \end{cases}$

B. $\begin{cases} E = A + 5 \\ A = S + 4 \end{cases}$

C. $\begin{cases} A = E + 5 \\ S = A - 4 \end{cases}$

D. $\begin{cases} E = A + 5 \\ A = S - 4 \end{cases}$

10) Right triangle ABC has two legs of lengths $4\ cm$ (AB) and $3\ cm$ (AC). What is the length of the third side (BC)?

A. $5\ cm$

B. $6\ cm$

C. $9\ cm$

D. $10\ cm$

11) Which is the longest time?

A. 24 hours
B. 1520 minutes
C. 3 days
D. 4200 seconds

12) A circle has a diameter of 10 inches. What is its approximate circumference?

A. 6.28
B. 25.12
C. 31.4
D. 35.12

13) Write 623 in expanded form, using exponents.

A. $(6 \times 10^3) + (2 \times 10^2) + (3 \times 10)$
B. $(6 \times 10^2) + (2 \times 10^1) - 5$
C. $(6 \times 10^2) + (2 \times 10^1) + 3$
D. $(6 \times 10^1) + (2 \times 10^2) + 3$

14) What is the area of an isosceles right triangle with hypotenuse that measures $8\ cm$?

A. $9\ cm$
B. $16\ cm$
C. $3\sqrt{2}\ cm$
D. $64\ cm$

15) A company pays its writer $5 for every 500 words written. How much will a writer earn for an article with 860 words?
 A. $12
 B. $5.6
 C. $8.6
 D. $10.7

16) A circular logo is enlarged to fit the lid of a jar. The new diameter is 20% larger than the original. By what percentage has the area of the logo increased?
 A. 20%
 B. 44%
 C. 69%
 D. 75%

17) $89.44 \div 0.05 = ?$
 A. 17.888
 B. 1,788.8
 C. 178.88
 D. 1.7888

18) What's the area of the non-shaded part of the following figure?
 A. 225
 B. 152
 C. 40
 D. 42

14   3

9

18

19) A bread recipe calls for $2\frac{1}{2}$ cups of flour. If you only have $1\frac{5}{4}$ cups, how much more flour is needed?

A. 1

B. $\frac{1}{2}$

C. 2

D. $\frac{1}{4}$

20) What is the maximum value for y if $y = -(x-2)^2 + 7$?

A. -7

B. -2

C. 2

D. 7

21) What is the solution of the following system of equations?
$$\begin{cases} -3x - y = -5 \\ 5x - 5y = 15 \end{cases}$$

A. $(-1, 2)$

B. $(2, -1)$

C. $(1, 4)$

D. $(4, -2)$

22) The equation of a line is given as: $y = 5x - 3$. Which of the following points does not lie on the line?

A. $(2, 7)$

B. $(-2, -13)$

C. $(4, 21)$

D. $(2, 7)$

23) The drivers at G & G trucking must report the mileage on their trucks each week. The mileage reading of Ed's vehicle was 41,905at the beginning of one week, and 42,054 at the end of the same week. What was the total number of miles driven by Ed that week?

A. 46 $miles$

B. 144 $miles$

C. 149 $miles$

D. 1,046 $miles$

24) What is the area of an isosceles right triangle that has one leg that measures $4\ cm$?

A. $8cm^2$

B. $36cm^2$

C. $3\sqrt{2cm^2}$

D. $72cm^2$

25) Which of the following is a factor of both $x^2 - 5x + 6$ and $x^2 - 6x + 8$?

A. $(x - 2)$

B. $(x + 4)$

C. $(x + 2)$

D. $(x - 4)$

26) $\begin{array}{r} 36\ \text{hr.}\ \ 38\ \text{min.} \\ -\ 23\ \text{hr.}\ \ 25\ \text{min.} \\ \hline \ \ \ \ \ \ \ \ \end{array}$

A. $12\ hr.\ 57\ min.$

B. $12\ hr.\ 47\ min.$

C. $13\ hr.\ 13\ min.$

D. $13\ hr.\ 57\ min.$

27) $\frac{14}{26}$ is equal to:

A. 5.3

B. 0.53

C. 0.05

D. 0.5

28) If $x + y = 10$, what is the value of $9x + 9y$?

A. 192

B. 104

C. 90

D. 48

29) What is the number of cubic feet of soil needed for a flower box 2 feet long, 10 inches wide, and 2 feet deep?

A. 22 cubic feet

B. 12 cubic feet

C. $\frac{10}{3}$ cubic feet

D. 2 cubic feet

30) A car uses 20 gallons of gas to travel 460 miles. How many miles per gallon does the car use?

A. 23 miles per gallon

B. 32 miles per gallon

C. 30 miles per gallon

D. 34 miles per gallon

31) What is the reciprocal of $\frac{x^3}{15}$?

A. $\frac{15}{x^3} - 1$

B. $\frac{48}{x^3}$

C. $\frac{15}{x^3} + 1$

D. $\frac{15}{x^3}$

32) Karen is 9 years older than her sister Michelle, and Michelle is 4 years younger than her brother David. If the sum of their ages is 91, how old is Michelle?

A. 21

B. 26

C. 28

D. 29

33) Mario loaned Jett $1,400 at a yearly interest rate of 6%. After one year what is the interest owned on this loan?

A. $1,260

B. $140

C. $84

D. $30

34) Calculate the area of a parallelogram with a base of 3 feet and height of 3.2 feet.

A. 2.8 square feet

B. 4.2 square feet

C. 5.8 square feet

D. 9.6 square feet

35) Ellis just got hired for on-the-road sales and will travel about 2,500 miles a week during an 90-hour work week. If the time spent traveling is $\frac{5}{3}$ of his week, how many hours a week will he be on the road?

A. Ellis spends about 34 hours of his 90-hour work week on the road.

B. Ellis spends about 40 hours of his 90-hour work week on the road.

C. Ellis spends about 48 hours of his 90-hour work week on the road.

D. Ellis spends about 150 hours of his 90-hour work week on the road.

36) Given that $x = 0.5$ and $y = 5$, what is the value of $2x^2(y + 4)$?

A. 4.5

B. 8.2

C. 12.2

D. 14.2

37) What is the area of the shaded region if the diameter of the bigger circle is 14 inches and the diameter of the smaller circle is 10 inches.

A. $16\,\pi\;inch^2$

B. $24\,\pi\;inch^2$

C. $36\,\pi\;inch^2$

D. $80\,\pi\;inch^2$

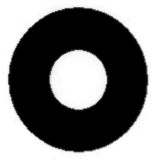

38) A shirt costing $500 is discounted 25%. After a month, the shirt is discounted another 15%. Which of the following expressions can be used to find the selling price of the shirt?

A. $(500)\,(0.70)$

B. $(500) - 500\,(0.30)$

C. $(500)\,(0.15) - (500)\,(0.15)$

D. $(500)\,(0.75)\,(0.85)$

39) A tree 40 feet tall casts a shadow 18 feet long. Jack is 5 feet tall. How long is Jack's shadow?

A. $2.25\;ft$

B. $4\;ft$

C. $5.25\;ft$

D. $7\;ft$

40) In a school, the ratio of number of boys to girls is $7:3$. If the number of boys is 210, what is the total number of students in the school?

A. 300

B. 500

C. 540

D. 600

41) If x is 35% percent of 620, what is x?

A. 185

B. 217

C. 402

D. 720

42) How many square feet of tile is needed for a $19\ feet\ x\ 19\ feet$ room?

A. 72 square feet

B. 108 square feet

C. 361 square feet

D. 416 square feet

43) $(4x + 4)\ (x + 5) =$

A. $4x + 8$

B. $4x + 3x + 15$

C. $4x^2 + 24x + 20$

D. $4x^2 + 3$

44) If $x \blacksquare y = \sqrt{x^2 + y}$, what is the value of $4 \blacksquare 9$?

A. $\sqrt{126}$

B. 6

C. 5

D. 4

45) There are three equal tanks of water. If $\frac{5}{2}$ of a tank contains 200 liters of water, what is the capacity of the three tanks of water together?

A. 1,400

B. 500

C. 240

D. 80

46) What is the result of the expression?

$$\begin{vmatrix} 3 & 6 \\ -1 & -3 \\ -5 & -1 \end{vmatrix} + \begin{vmatrix} 2 & -1 \\ 6 & 4 \\ 1 & 3 \end{vmatrix}$$

A. $\begin{vmatrix} 1 & -1 \\ 6 & 0 \\ 2 & 3 \end{vmatrix}$

B. $\begin{vmatrix} 3 & 7 \\ -1 & -3 \\ -5 & -1 \end{vmatrix}$

C. $\begin{vmatrix} 5 & 5 \\ 5 & 1 \\ -4 & 2 \end{vmatrix}$

D. $\begin{vmatrix} 5 & -3 \\ -6 & 1 \\ -10 & -3 \end{vmatrix}$

47) The average weight of 20 girls in a class is $55 \, kg$ and the average weight of 35 boys in the same class is $70kg$. What is the average weight of all the 55 students in that class?

A. 60

B. 61.28

C. 64.54

D. 65.9

IF YOU FINISH BEFORE TIME IS CALLED, YOU MAY CHECK YOUR WORK ON THIS SECTION. STOP

ISEE Upper Level Math Practice Test Answers and Explanations

※ Now, it's time to review your results to see where you went wrong and what areas you need to improve!

ISEE Upper Level Math Practice Test 2 Answer Key											
Quantitative Reasoning						Mathematics Achievement					
1	B	17	A	33	C	1	B	17	A	33	D
2	C	18	D	34	A	2	A	18	B	34	A
3	C	19	B	35	A	3	B	19	D	35	B
4	D	20	B	36	A	4	D	20	C	36	D
5	A	21	D	37	D	5	D	21	C	37	D
6	D	22	A			6	D	22	D	38	A
7	A	23	B			7	A	23	B	39	A
8	C	24	D			8	B	24	A	40	B
9	C	25	A			9	A	25	A	41	C
10	D	26	D			10	C	26	B	42	C
11	C	27	D			11	C	27	C	43	C
12	D	28	C			12	C	28	C	44	C
13	D	29	A			13	C	29	A	45	C
14	D	30	A			14	B	30	C	46	C
15	B	31	B			15	C	31	D	47	B
16	D	32	B			16	B	32	C		

ISEE UPPER LEVEL Math Practice Test Section 1

Answers and Explanations

1) Choice C is correct

$(4x + 9) - (4x - 3) = 4x - 4x + 9 + 3 = 12$

2) Choice B is correct

Find the value of each choice:
$2 \times 2 \times 5 \times 5 = 100$
$2 \times 2 \times 2 \times 5 \times 5 \times 7 = 1400$
$2 \times 7 = 14$
$2 \times 2 \times 2 \times 5 \times 7 = 280$

3) Choice D is correct

Write a proportion and solve. $\frac{6 in}{150 feet} = \frac{20 in}{x} \rightarrow x = \frac{150 \times 20}{6} = 500$

4) Choice A is correct

Let x be all expenses, then $\frac{22}{100} x = \$660 \rightarrow x = \frac{100 \times \$550}{22} = \$2500$
He spent for his rent: $\frac{27}{100} \times \$2500 = \$675$

5) Choice A is correct

Area of the circle is less than $14\,\pi$. Use the formula of areas of circles.
$Area = \pi r^2 \Rightarrow 49\,\pi > \pi r^2 \Rightarrow 49 > r^2 \Rightarrow r < 7$
Radius of the circle is less than 7. Let's put 7 for the radius. Now, use the circumference formula: $Circumference = 2\pi r = 2\pi\,(7) = 14\,\pi$
Since the radius of the circle is less than 7. Then, the circumference of the circle must be less than $14\,\pi$. Only choice A is less than $14\,\pi$

6) Choice C is correct

Recall that the formula for the average is: $Average = \frac{sum\ of\ data}{number\ of\ data}$

First, compute the total weight of all balls in the basket: $35g = \frac{total\ wei}{25\ balls}$
$35g \times 25 = total\ weight = 875g$
Next, find the total weight of the 5 largest marbles: $50g = \frac{total\ weight}{5\ marbles}$
$50\,g \times 5 = total\ weight = 250\,g$
The total weight of the heaviest balls is $250\,g$. Then, the total weight of the remaining 20 balls is $625g$. $875\,g - 250\,g = 625\,g$.
The average weight of the remaining balls: $Average = \frac{625\,g}{20\ marbles} = 31.25g$ per ball

230

7) Choice C is correct

The smallest possible value of $f(x)$ will occur when $x = 0$. Since x^2 is always positive, any positive or negative value of x will make the value of $f(x)$ greater than 6. Substitute 0 for x and evaluate the expression: $f(0) = (0)^2 + 6 = 6$

8) Choice D is correct

There is not enough information to determine the answer of the question. An average speed represents a distance divided by time and it does not provide information about the speed at specific time. Alice could drove exactly 45 miles per hour from start to finish, or she could drive 65 miles per hour for half of distance and 45 miles per hour for the other half.

9) Choice D is correct

There are 2 sets of values, one set from 1 to n, and the other set from $n + 1$ to $2n$. Since the second set begins immediately after the first set, the two sets can be combined. The sum of the positive integers from 1 to $2n$ inclusive is equal to the sum of the positive integers from 1 to n plus the sum of the positive integers from $n + 1$ to $2n$: $3,350 + 4,866 = 8216$

10) Choice C is correct

The percentage discount is the reduction in price divided by the original price. The difference between original price and sale price is: $\$14.65 - \$8.34 = \$6.31$

The percentage discount is this difference divided by the original price: $\$6.31 \div \$14.65 \cong 0.43$

Convert the decimal to a percentage by multiplying by 100%: $0.43 \times 100\% = 43\%$

11) Choice C is correct

Let's review the choices provided:

A. Number of books sold in April is: 380

Number of books sold in July is: $760 \rightarrow \frac{380}{760} = \frac{38}{76} = \frac{1}{2}$

B. number of books sold in July is: 760

Half the number of books sold in May is: $\frac{1140}{2} = 570 \rightarrow 760 > 570$

C. number of books sold in June is: 190

Half the number of books sold in April is: $\frac{380}{2} = 190 \rightarrow 190 = 190$

D. $380 + 190 = 570 < 760$

Only choice C is correct.

12) Choice D is correct

The median of a set of data is the value located in the middle of the data set. Combine the 2 sets provided, and organize them in ascending order: $\{2, 4, 5, 7, 9, 11, 13, 15, 16, 18\}$

Since there are an even number of items in the resulting list, the median is the average of the two middle numbers. $Median = (9 + 11) \div 2 = 10$

13) Choice D is correct

If 17 balls are removed from the bag at random, there will be one ball in the bag.

The probability of choosing a brown ball is 1 out of 19. Therefore, the probability of not choosing a brown ball is 17 out of 19 and the probability of having not a brown ball after removing 17 balls is the same.

14) Choice D is correct

Let x be the number of current stamps in the collection. Then:

$\frac{5}{6}x - x = 100 \rightarrow \frac{1}{6}x = 100 \rightarrow x = 600$

50% more of 600 is: $600 + 0.50 \times 600 = 600 + 300 = 900$

15) Choice B is correct

$(x^2 - y^2) = (x - y)(x + y)$, Then: $x^2 - y^2 = 7 \times 6 = 42$

16) Choice A is correct

The formula for the area of a rectangle is: $Area = Width \times Length$

It is given that $L = 3W$ and that $A = 100$. Substitute the given values into our equation and solve for W: $100 = w \times 3w \rightarrow 100 = 4w^2 \rightarrow w^2 = 25 \rightarrow w = 5$

It is given that $L = 3W$, therefore, $L = 3 \times 5 = 15$

The perimeter of a rectangle is: $2L + 2W$

Perimeter $= 2 \times 15 + 2 \times 5$,　　　Perimeter $= 40$

17) Choice D is correct

The ratio of boy to girls is $7: 4$. Therefore, there are 7 boys out of 11 students. To find the answer, first divide the total number of students by 11, then multiply the result by 7.

$55 \div 11 = 5 \Rightarrow 5 \times 7 = 35$

There are 35 boys and 20 (55 – 35) girls. So, 15 more girls should be enrolled to make the ratio $1: 1$

18) Choice D is correct

The sum of 8 numbers is greater than 320 and less than 480. Then, the average of the 8 numbers must be greater than 40 and less than 60.

$\frac{320}{8} < x < \frac{480}{8} \rightarrow 40 < x < 60$

The only choice that is between 40 and 60 is 45.

19) Choice D is correct

Let x be number of gallons the tank can hold when it is full. Then: $\frac{5}{2}x = 35 \rightarrow x = \frac{2}{5} \times 35 = 14$

20) Choice B is correct

If two triangles are similar, then the ratios of corresponding sides are equal.

$\frac{AC}{AE} = \frac{BC}{DE} = \frac{20}{10} = 2, \frac{AC}{AE} = 2$

This ratio can be used to find the length of AC: $AC = 2 \times AE$

$AC = 2 \times 10 \rightarrow AC = 20$

The length of AE is given as 10 and we now know the length of AC is 20, therefore:

$EC = AC - AE$

$EC = 20 - 10$,　　　$EC = 10$

21) Choice B is correct

In order to solve for the variable b, first take square roots on both sides:

$\sqrt{a} = \frac{cz}{b}$, then multiply both sides by b: $b\sqrt{a} = cz$

Now, divide both sides by $\sqrt{a}$: $b = \frac{cz}{\sqrt{a}}$

22) Choice C is correct

Column A: $2^2 = 4$, Column B: $\sqrt[3]{64} = 4$ (recall that $4^3 = 16$)

23) Choice A is correct

A number raised to the exponent $(\frac{1}{2})$ is the same thing as evaluating the square root of the number. Therefore: $(59)^{\frac{1}{2}} = \sqrt{59}$

Since $\sqrt{64}$ is greater than $\sqrt{59}$, column A ($\sqrt{64} = 8$) is greater than $\sqrt{59}$.

24) Choice A is correct

The average is the sum of all terms divided by the number of terms.

$15 + 25 + 27 = 67$, $67 \div 3 = 22.33$

This is greater than 20.

25) Choice B is correct

Since both columns have 675 as a factor, we can ignore that number.

$11 \times 24 = 264$, $18 \times 17 = 306$

Column B is greater

26) Choice D is correct

Since x is an integer and can be positive and negative, then the relationship cannot be determined from the information given. Let's choose some values for x.

$x = 1$, then the value in column A is greater. $1 > \frac{1}{-2}$

Let's choose a negative value for x. $x = -1$, then the value in column B is greater.

$-1 < \frac{-1}{-2} \rightarrow -1 < \frac{1}{2}$

27) Choice C is correct

To raise a quantity to a negative power, invert the numerator and denominator, and then raise the base to the indicated power. Therefore:

$(\frac{5}{1})^{-2} = (\frac{1}{5})^2$, The Columns are the same value.

28) Choice A is correct

First, simplify the inequality: $x + 2 > 5x \rightarrow 2 > 4x \rightarrow \frac{2}{4} > x \rightarrow \frac{1}{2} > x$

Since x is less than $\frac{1}{2}$, and x can be 0 (greater than -1) or -2 (less than -1), the relationship cannot be determined.

29) Choice A is correct

First, find the values of x in both columns.

Column A: $6|2x - 4| = 6 \rightarrow |2x - 4| = 1$

$2x - 4$ can be 1 or -1.

$2x - 4 = 1 \rightarrow 2x = 5 \rightarrow x = \dfrac{5}{2}$

$2x - 4 = -1 \rightarrow 2x = 3 \rightarrow x = \dfrac{3}{2}$

Column B: $6|2x + 4| = 6 \rightarrow |2x + 4| = 1$

$2x + 4$ can be 1 or -1.

$2x + 4 = 1 \rightarrow 2x = -3 \rightarrow x = -\dfrac{3}{2}$

$2x + 4 = -1 \rightarrow 2x = -5 \rightarrow x = -\dfrac{5}{2}$

The greatest value of x in column A is $\dfrac{5}{2}$ and the greatest value of x in column B is $-\dfrac{3}{2}$.

30) Choice D is correct

Simplify both columns.

Column A: $(x)^3(x)^4 = x^7$

Column B: $(x^3)^4 = x^{12}$

Column A evaluates to x^7 and Column B evaluates to x^{12}. In the case where $x = 0$, the two columns will be equal, but if $x = 2$, the two columns will not be equal. Consequently, the relationship cannot be determined.

31) Choice D is correct

The probability that an event will occur + the probability that that event will NOT occur must equal 1. Since we don't have any numerical information about the probability, it is possible that the probability that event x occurs is 25%, 50% or any other percent. The probability that event x will not occur will always be 100% minus the probability that event x does occur. Because both columns can exhibit a range of values, the relationship cannot be determined.

32) Choice A is correct

Let x be the original price of the sport jacket. The selling price of a sport jacket including 25% discount is \$51. Then: $x - 0.25x = 51 \rightarrow 0.75x = 51 \rightarrow x = \dfrac{51}{0.75} = 68$

The original price of the jacket is \$68 which is greater than column B (\$67).

33) Choice D is correct

Factor the expression if possible. Begin by moving all terms to one side before factoring:

$x^2 - 5x - 8 = 6$

$x^2 - 5x - 14 = 0$

To factor this quadratic, find two numbers that multiply to -14 and sum to -5:

$(x - 7)(x + 2) = 0$

Set each expression in parentheses equal to 0 and solve: $x - 7 = 0$

$x = 7$
$x + 2 = 0$
$x = -2$
Quadratic equations can have TWO possible solutions. Since one of these is greater than 0 and one of them is less than 0, we cannot determine the relationship between the columns.

34) Choice B is correct

The value of x has to be less than 50, which is less than 60. Column B is greater.
Recall that when the positive powers of numbers between 0 and 1 increases, the value of the number decrease. For example: $(0.5)^2 > (0.5)^3 \rightarrow 0.25 > 0.125$
So, $(0.69)^{33} > (0.69)^{34}$

35) Choice B is correct

Because of the word "and" the events described in each column must be calculated separately and then multiplied: For column A: Probability of rolling a 6: $\frac{1}{6}$
Probability of getting heads: $\frac{1}{2}$
$\frac{1}{6} \times \frac{1}{12} = \frac{1}{12}$
For column B: Probability of an even number: $\frac{3}{6} = \frac{1}{2}$
Probability of getting a spade: $\frac{13}{52} = \frac{1}{4}$
$\frac{1}{2} \times \frac{1}{4} = \frac{1}{8}$
Since $\frac{1}{8}$ is a larger number than $\frac{1}{12}$, Colum B is greater

36) Choice A is correct

Sum of one quarter, two nickels, and three pennies is: $\$0.25 + 2(\$0.05) + \$0.03 = \$0.38$

37) Choice A is correct

Let's consider the properties of odd and even integers:
Odd +/− Odd = Even
Even +/− Even = Even
Odd +/− Even = Odd
Odd × Odd = Odd
Even × Even = Even
Odd × Even = Even
Now let's review the columns. For column A: $(x + y)^2 - y$
(odd + even)² − even
(odd)² − even
(odd)(odd) − even
odd − even
odd
For Column B:

$(y)(x - y)$
(even)(odd – even)
(even)(odd)
even
Since an odd number is considered greater according to the problem statement, the answer is A.

ISEE UPPER LEVEL Math Practice Test Section 2

Answers and Explanations

1) Choice D is correct.

Plug in each pair of numbers in the equation. The answer should be 20.

A. $(2, 1)$: $4(2) + 6(1) = 14$ No!
B. $(-1, 3)$: $4(-1) + 6(2) = 8$ No!
C. $(-2, 2)$: $4(-2) + 6(2) = 4$ No!
D. $(2, 2)$: $4(2) + 6(2) = 20$ Yes!

  2) Choice D is correct

Let x be the integer. Then: $2x - 5 = 91$, Add 5 both sides: $2x = 96$
Divide both sides by 2: $x = 48$

3) Choice A is correct

First, multiply both sides of inequality by 5. Then: $\frac{|3+x|}{5} \leq 8 \rightarrow |3 + x| \leq 40$
Since $3 + x$ can be positive or negative, then: $3 + x \leq 40 \ or \ 3 + x \geq -40$
Then: $x \leq 37 \ or \ x \geq -43$

4) Choice B is correct

Subtract $\frac{1}{5b}$ and $\frac{1}{b^2}$ from both sides of the equation. Then: $\frac{1}{5b^2} + \frac{1}{5b} = \frac{1}{b^2} \rightarrow \frac{1}{5b^2} - \frac{1}{b^2} = -\frac{1}{5b}$
Multiply both numerator and denominator of the fraction $\frac{1}{b^2}$ by 5. Then: $\frac{1}{5b^2} - \frac{5}{5b^2} = -\frac{1}{5b}$
Simplify the first side of the equation: $-\frac{4}{5b^2} = -\frac{1}{5b}$
Use cross multiplication method: $20b = 5b^2 \rightarrow 20 = 5b \rightarrow b = 4$

5) Choice A is correct

The sum of supplement angles is 180. Let x be that angle. Therefore, $x + 5x = 180$
$6x = 180$, divide both sides by 6: $x = 30$

6) Choice B is correct

$x\% \ 26 = 1.3 \rightarrow \frac{x}{100} 26 = 1.3 \rightarrow \ x = \frac{1.3 \times 100}{26} = 5$

7) Choice B is correct

Plug in the value of $x = 20$ into both equations. Then:
$C(x) = x^2 + 2x = (20)^2 + 2(20) = 400 + 40 = 440$
$R(x) = 40x = 40 \times 20 = 800$
$800 - 440 = 360$
So, the profit is $360.

8) Choice C is correct

$7x^3y^3(2x^3y)^3 = 7x^3y^3(8x^9y^3) = 56x^{12}y^6$

9) Choice D is correct

From choices provided, only choice D is correct.
$E = 5 + A$
$A = S - 4$

10) Choice A is correct

Use Pythagorean Theorem: $a^2 + b^2 = c^2 \Rightarrow 4^2 + 3^2 = c^2 \Rightarrow 25 = c^2 \Rightarrow c = 5$

11) Choice C is correct

$24\ hours = 86{,}400\ seconds$, $1520\ minutes = 91{,}200\ seconds$
$3\ days = 72\ hours = 259{,}200\ seconds$

12) Choice C is correct

$C = 2\pi r \Rightarrow C = 2\pi \times 5 = 10\pi \Rightarrow \pi = 3.14 \rightarrow C = 10\pi = 31.4$

13) Choice C is correct

Let's review the choices provided:
A. $(6 \times 10^3) + (2 \times 10^2) + (3 \times 10) = 6{,}000 + 200 + 30 = 6{,}230$
B. $(6 \times 10^2) + (2 \times 10^1) - 5 = 600 + 20 - 5 = 615$
C. $(6 \times 10^2) + (2 \times 10^1) + 3 = 600 + 20 + 3 = 623$
D. $(6 \times 10^1) + (2 \times 10^2) + 3 = 60 + 20 + 3 = 83$
Only choice C equals to 623.

14) Choice B is correct

First draw an isosceles triangle. Remember that two sides of the triangle are equal.
Let put a for the legs. Then: Use Pythagorean theorem to find the value of a:
$a^2 + b^2 = c^2 \rightarrow a^2 + a^2 = 8^2$
Simplify: $2a^2 = 64 \rightarrow a^2 = 32 \rightarrow a = \sqrt{32}$

Isosceles right triangle

$a = \sqrt{32} \Rightarrow$ area of the triangle is $= \frac{1}{2}\left(\sqrt{32} \times \sqrt{32}\right) = \frac{1}{2} \times 32 = 16\ cm^2$

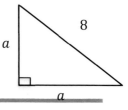

15) Choice C is correct

$$\frac{5}{500} = \frac{x}{860} \Rightarrow x = \frac{5 \times 860}{500} = 8.6$$

16) Choice B is correct

Area of a circle equals: $A = \pi r^2$
The new diameter is 20% larger than the original then the new radius is also 20% larger than the original. 20% larger than r is $1.2r$. Then, the area of larger circle is:
$A = \pi r^2 = \pi (1.2r)^2 = \pi(1.44r^2) = 1.44\pi r^2$
$1.44\pi r^2$ is 44% bigger than πr^2.

17) Choice B is correct

$89.44 \div 0.05 = 1,788.8$

18) Choice A is correct

The area of the non-shaded region is equal to the area of the bigger rectangle subtracted by the area of smaller rectangle.
Area of the bigger rectangle = $14 \times 18 = 252$
Area of the smaller rectangle = $9 \times 3 = 27$
Area of the non-shaded region = $252 - 27 = 225$

19) Choice D is correct

$2\frac{1}{2} - 1\frac{5}{4} =$, Break off 1 from 2: $2\frac{1}{2} = 1\frac{3}{2}$
$1\frac{3}{2} - 1\frac{5}{4} =$
Subtract whole numbers: $1 - 1 = 0$, Combine fractions: $\frac{3}{2} - \frac{5}{4} = \frac{1}{4}$

20) Choice D is correct

To find the maximum value of y, the expression $(x - 2)^2$ must be equal to 0. Because it has a negative sign. Since $x - 2$ is to the power of 2, it cannot be negative. To get 0 for the expression $(x - 2)^2$, x must be 2.
Plug in 2 for x in the equation: $y = -(x - 2)^2 + 6 \rightarrow y = -(2 - 2)^2 + 7 = 7$
The maximum value of y is 7.

21) Choice B is correct

$\begin{cases} -3x - y = -5 \\ 5x - 5y = 15 \end{cases} \Rightarrow$ Multiplication (-5) in first equation $\Rightarrow \begin{cases} 15 + 5y = 25 \\ 5x - 5y = 15 \end{cases}$
Add two equations together $\Rightarrow 20x = 40 \Rightarrow x = 2$ then: $y = -1$

22) Choice C is correct

Let's review the choices provided. Put the values of x and y in the equation.

A. $(2, 7)$ $\Rightarrow$ $x = 1 \Rightarrow y = 7$ This is true!
B. $(-2, -13)$ $\Rightarrow x = -2 \Rightarrow y = -13$ This is true!
C. $(4, 21)$ $\Rightarrow x = 4 \Rightarrow y = 17$ This is not true!
D. $(2, 7)$ $\Rightarrow x = 2 \Rightarrow y = 7$ This is true!

23) Choice C is correct

To find total number of miles driven by Ed that week, you only need to subtract 42,054 from 41,905.
$42,054 - 41,905 = 149$

24) Choice A is correct

First draw an isosceles triangle. Remember that two sides of the triangle are equal.

Isosceles right triangle

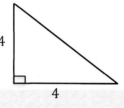

Let put a for the legs. Then:
$a = 4 \Rightarrow$ area of the triangle is $= \frac{1}{2}(4 \times 4) = \frac{16}{2} = 8 \ cm^2$

25) Choice A is correct

Factor each trinomial $x^2 - 5x + 6$ and $x^2 - 6x + 8$
$x^2 - 5x + 6 \Rightarrow (x - 2)(x - 3)$
$x^2 - 6x + 8 \Rightarrow (x - 2)(x - 4)$
The common factor of both expressions is $(x - 2)$.

26) Choice C is correct

$\quad$ 36 hr. 38 min.
$\underline{- \ 23 \ \text{hr.} 25 \ \text{min.}}$
13 hr *and* 13min

27) Choice B is correct

$\frac{14}{26} = 0.53$

28) Choice C is correct

$x + y = 10$
Then: $9x + 9y = 9(x + y) = 9 \times 10 = 90$

29) Choice C is correct

First, convert all measurement to foot.

One foot is 12 inches. Then: 12 inches $= \frac{10}{12} = \frac{5}{6}$ feet

The volume flower box is: length × width × height $= 2 \times \frac{5}{6} \times 2 = \frac{10}{3}$ cubic feet.

30) Choice A is correct

$\frac{460}{20} = 23$

31) Choice D is correct

$\frac{x^3}{15} \Rightarrow$ reciprocal is : $\frac{15}{x^3}$

32) Choice B is correct

Let's write equations based on the information provided:
$Michelle = Karen - 9$
$Michelle = David - 4$
$Karen + Michelle + David = 82$
$Karen - 9 = Michelle \Rightarrow Karen = Michelle + 9$
$Karen + Michelle + David = 91$
Now, replace the ages of Karen and David by Michelle. Then:
$Michelle + 9 + Michelle + Michelle + 4 = 91$
$3Michelle + 13 = 91 \Rightarrow 3Michelle = 91 - 13$
$3Michelle = 78$
$Michelle = 26$

33) Choice C is correct

Use interest rate formula: $Interest = principal \times rate \times time = 1,400 \times 0.06 \times 1 = 84$

34) Choice D is correct

$A = bh, \qquad A = 3 \times 3.2 = 9.6$

35) Choice D is correct

Ellis travels $\frac{5}{3}$ of 90 hours. $\frac{5}{3} \times 90 = 150$, Ellis will be on the road for 150 hours.

36) Choice A is correct

Plug in the values of x and y in the expression:
$2x^2(y + 4) = 2(0.5)^2(5 + 4) = 2(0.25)(9) = 4.5$

37) Choice B is correct.

To find the area of the shaded region subtract smaller circle from bigger circle.
$S_{bigger} - S_{smaller} = \pi(r_{bigger})^2 - \pi(r_{smaller})^2 \Rightarrow S_{bigger} - S_{smaller} = \pi(7)^2 - \pi(5)^2 \Rightarrow$
$49\pi - 25\pi = 24\pi$

38) Choice D is correct

To find the discount, multiply the number by $(100\% - rate\ of\ discount)$.

Therefore, for the first discount we get: $(500)(100\% - 25\%) = (500)(0.75)$
For the next 15% discount: $(500)(0.75)(0.85)$

39) Choice A is correct

Write a proportion and solve for the missing number. $\frac{40}{18} = \frac{5}{x} \rightarrow 40x = 18 \times 5 = 90$

$40x = 90 \rightarrow x = \dfrac{90}{40} = 2.25$

40) Choice A is correct

The ratio of boys to girls is $7:3$. Therefore, there are 7 boys out of 10 students. To find the answer, first divide the number of boys by 7, then multiply the result by 10.
$210 \div 7 = 30 \Rightarrow 30 \times 10 = 300$

41) Choice B is correct

$\dfrac{35}{100} \times 620 = x \rightarrow x = 217$

42) Choice C is correct
The area of a $19\ feet\ x\ 19\ feet$ room is 361 square feet. $19 \times 19 = 361$

43) Choice C is correct

Use FOIL (First, Out, In, Last)
$(4x + 4)(x + 5) = 4x^2 + 20x + 4x + 20 = 4x^2 + 24x + 20$

44) Choice C is correct

Plug in the values of x and y in the equation: $4 \blacksquare 9 = \sqrt{4^2 + 9} = \sqrt{16 + 9} = \sqrt{25} = 5$

45) Choice C is correct

Let x be the capacity of one tank. Then, $\frac{5}{2}x = 200 \rightarrow x = \frac{200 \times 2}{5} = 80$ Liters
The amount of water in three tanks is equal to: $3 \times 80 = 240$ Liters

46) Choice C is correct.
To add two matrices, first we need to find corresponding members from each matrix.

$$\begin{vmatrix} 3 & 6 \\ -1 & -3 \\ -5 & -1 \end{vmatrix} + \begin{vmatrix} 2 & -1 \\ 6 & 4 \\ 1 & 3 \end{vmatrix} = \begin{vmatrix} 5 & 5 \\ 5 & 1 \\ -4 & 2 \end{vmatrix}$$

47) Choice C is correct

$Average = \dfrac{\text{sum of terms}}{\text{number of terms}}$

The sum of the weight of all girls is: $20 \times 55 = 1100\ kg$
The sum of the weight of all boys is: $35 \times 70 = 2450\ kg$
The sum of the weight of all students is: $1100 + 2450 = 3550\ kg$

Average $= \frac{3550}{55} = 64.54$

"Effortless Math Education" Publications

Effortless Math authors' team strives to prepare and publish the best quality ISEE Upper Level Mathematics learning resources to make learning Math easier for all. We hope that our publications help you learn Math in an effective way and prepare for the ISEE Upper Level test.

We all in Effortless Math wish you good luck and successful studies!

Effortless Math Authors

www.EffortlessMath.com

... So Much More Online!

✓ FREE Math lessons

✓ More Math learning books!

✓ Mathematics Worksheets

✓ Online Math Tutors

Need a PDF version of this book?

Visit www.EffortlessMath.com